›nds in Mathematics

's in Mathematics is a series devoted to the publication of volumes arising conferences and lecture series focusing on a particular topic from any area of ›matics. Its aim is to make current developments available to the community as y as possible without compromise to quality and to archive these for reference.

›als for volumes can be submitted using the Online Book Project Submission ›t our website www.birkhauser-science.com.

›l submitted for publication must be screened and prepared as follows:

›tributions should undergo a reviewing process similar to that carried out by › and be checked for correct use of language which, as a rule, is English. › without proofs, or which do not contain any significantly new results, ›e rejected. High quality survey papers, however, are welcome.

›ct the organizers to deliver manuscripts in a form that is essentially ready ›t reproduction. Any version of TEX is acceptable, but the entire collection ›nust be in one particular dialect of TEX and unified according to simple ›ns available from Birkhäuser.

›ore, in order to guarantee the timely appearance of the proceedings it is ›that the final version of the entire material be submitted no later than one › the conference.

Tr

Tren
from
math
rapid

Propo
Form

Materi

All co
journal
Article
should

We exp
for dire
of files
instructi

Furthern
essential
year afte

Joaquim Duran i Lamiel • Alberto Maione
Editors

New Frontiers in Homogenization and Fractional Calculus

Birkhäuser

Editors

Joaquim Duran i Lamiel
Campus de Bellaterra
Centre de Recerca Matemàtica
Bellaterra, Spain

Alberto Maione
Dipartimento di Matematica, Politecnico di Milano
Campus Leonardo, Edificio 14 "Nave"
Milano, Italy

ISSN 2297-0215 ISSN 2297-024X (electronic)
Trends in Mathematics
ISBN 978-3-032-20097-6 ISBN 978-3-032-20098-3 (eBook)
https://doi.org/10.1007/978-3-032-20098-3

Mathematics Subject Classification: 35R11, 35A15, 35B27, 49J45, 47A10

This book is published under the imprint Birkhäuser, www.birkhauser-science.com by the registered company Springer Nature Switzerland AG
The registered company address is: Gewerbestrasse 11, 6330 Cham, Switzerland

Preface

This volume originates from the conference *New Frontiers in Homogenization and Fractional Calculus*, held at the Centre de Recerca Matemàtica (CRM), Barcelona, on March 24–25, 2025. The event was organized to celebrate the 50th anniversary of the mathematical technique of Γ-convergence, introduced by Ennio De Giorgi and Tullio Franzoni in 1975. Originally developed for applied purposes, this technique has become a fundamental tool in the study of a wide variety of scientific phenomena, including homogenization, phase transitions, and the asymptotic analysis of partial differential equations. In parallel, there has been a growing interest in nonlocal problems, due to their relevance in probability theory and their increasing relevance in material science. Besides being an intrinsically challenging area of mathematics, this theory plays an essential role in modeling nonlocal phenomena.

The majority of the conference speakers kindly agreed to contribute to this volume. Their chapters either present an extended description of the results discussed at the conference or address related topics that have not yet been extensively published in the literature or are particularly demanding to present to a broader audience. The contributions span a wide range of themes, chosen to reflect the main research directions active in the field. Moreover, since these techniques extend beyond pure analytical theory, the volume also includes works with a numerical perspective. The book is aimed at a broad audience: from students in mathematics, physics, and engineering, to researchers and experts in these areas.

The Volume Editors wish to express their sincere gratitude to Xavier Cabré, ICREA-UPC-CRM, for supporting these initiatives, and to Dorothy Mazlum, Senior Editor in Mathematics at Birkhäuser, for suggesting the publication of this conference book and for her constant assistance throughout its preparation. The editors are also indebted to the authors of the chapters, as well as to all the conference speakers and participants, who made the event both scientifically valuable and truly inspiring. A warm thanks goes to the CRM staff, in particular to Vanessa Ramirez and Maria Batlló, for their unfailing assistance, without which the event would not have been possible.

Bellaterra, Spain
Milano, Italy
21 January 2026

Joaquim Duran
Alberto Maione

Contents

Contributors

Andrea Braides Department of Mathematics, University of Rome Tor Vergata, Rome, Italy

Maicol Caponi Dipartimento di Ingegneria e Scienze dell'Informazione e Matematica, Università degli Studi dell'Aquila, L'Aquila, Italy

Alessandro Carbotti Dipartimento di Matematica e Fisica "Ennio De Giorgi", Università del Salento, Lecce, Italy

Patrick Dondl Abteilung für Angewandte Mathematik, Albert-Ludwigs-Universität Freiburg, Freiburg im Breisgau, Germany

Joaquim Duran Centre de Recerca Matemàtica, Bellaterra, Spain

Alberto Maione Dipartimento di Matematica, Politecnico di Milano, Campus Leonardo, Edificio 14 "Nave", Milano, Italy

Fabio Paronetto Department of Mathematics "Tullio Levi-Civita", Università di Padova, Padova, Italy

Ludwig Striet Abteilung für Angewandte Mathematik, Albert-Ludwigs-Universität Freiburg, Freiburg im Breisgau, Germany

Simone Verzellesi Dipartimento di Matematica "Tullio Levi-Civita", Università degli Studi di Padova, Padova, Italy

Singular-Perturbation Problems in Fractional Sobolev Spaces—Recent Results

Andrea Braides

Abstract Singular perturbations have been used to select solutions of (non-convex) variational problems with a multiplicity of minimizers. The prototype of such an approach is the gradient theory of phase transitions by L. Modica, who specialized some earlier Gamma-convergence results by himself and S. Mortola contained in a seminal paper, validating the so-called minimal-interface criterion. I will give an overview of some recent results on perturbations with fractional and higher-order seminorms both in the framework of phase transitions and of free-discontinuity problems, relating these results with the Bourgain-Brezis-Mironescu and Maz'ya-Shaposhnikova limit analysis for fractional Sobolev seminorms, and with the theory of Gamma-expansions.

1 Introduction

With the aim of introducing singular-perturbation problems as a part of the development of Γ-convergence, I begin with some personal recollections about the early times of Γ-convergence, for which those were a prototypical issue. When I started my degree thesis work under the guidance of Ennio De Giorgi in 1981, the theoretical foundations of Γ-convergence had been already laid and it was already a stable theory. I got my personal idea of what had been the progress of this theory in the preceding years from those who directly worked with De Giorgi at the first developments, from Tullio Franzoni, to Sergio Spagnolo, to Luciano Modica, who was the first to give a course on Γ-convergence at Pisa University in the academic year 1982–1983.

An example that De Giorgi liked to quote was (one-dimensional) elliptic homogenization, with an oscillating coefficient taking two values (he called it the 'Greek function', as he would freely associate its graph to an hellenic ornamental motive).

A. Braides (✉)
Department of Mathematics, University of Rome Tor Vergata, Rome, Italy
e-mail: braides@mat.uniroma2.it

J. Duran i Lamiel, A. Maione (eds.), *New Frontiers in Homogenization and Fractional Calculus*, Trends in Mathematics,
https://doi.org/10.1007/978-3-032-20098-3_1

Indeed, Γ-convergence had directly stemmed from the interest in homogenization problems for elliptic operators, that De Giorgi and Spagnolo interpreted as a convergence of the energies. Even though Γ-convergence was then born as a variational counterpart of G-convergence, the convergence of Green's operators for elliptic problems, it was clear to De Giorgi that the flexibility of a purely topological convergence made the new notion not restricted to a specific class of problems or structure of the functionals (as had been Mosco convergence for convex functionals). Beside the theoretical developments leading to the paper in collaboration with Franzoni and the successive efforts to explore all possible variants of the convergence, De Giorgi was keen on obtaining results showing how Γ-convergence was not restricted to the usual applications to homogenization, since it needed no regularity, convexity or even coerciveness assumptions. I recall the early paper by Buttazzo and Dal Maso [13], where an example is given of convergence of non-convex and non-equi-Lipschitz integral functionals with a loss of smoothness; also my thesis work was about the development of a homogenization theory for non-coercive energies [6]. In this context the work of Modica and Mortola [23] was aimed at showing how the type of functional obtained as a Γ-limit may be completely different from the converging ones. The functionals they examined have the form

$$F_\varepsilon(u) = \frac{1}{\varepsilon}\int_\Omega W\Big(\frac{u}{\varepsilon}\Big)dx + \varepsilon\int_\Omega |\nabla u|^2 dx. \tag{1}$$

They made the explicit choice $W(z) = \sin^2(\pi z)$, but their result is valid for any 1-periodic non-negative continuous function W vanishing only on $\mathbb{Z}$. The corresponding Γ-limit in the L^1-topology is then simply

$$F_0(u) = m_W \|Du\|(\Omega), \qquad m_W = 2\int_0^1 \sqrt{W(s)}\,ds \tag{2}$$

with domain $BV(\Omega)$ and with $\|Du\|(\Omega)$ the total variation of Du on Ω. A second result in the Modica and Mortola paper concerns the functionals

$$F_\varepsilon(u) = \frac{1}{\varepsilon}\int_\Omega W(u)dx + \varepsilon\int_\Omega |\nabla u|^2 dx, \tag{3}$$

which differ from the first ones in the first term, now forcing in the limit u to take integer values. The Γ-limit has the same form of the first one, but with domain $BV(\Omega;\mathbb{Z})$, the set of BV functions taking values in $\mathbb{Z}$. Even though Modica and Mortola justify their analysis with possible applications to minimal surfaces, the application of their (second) result to phase-transition problems had to wait for almost a decade.

The result of Modica and Mortola can be seen as a singular-perturbation analysis. Singular perturbations are often used to obtain approximate solutions to problems which possess no solution or infinitely many ones, or to give a selection criterion among different solutions. One can think of the theory of viscosity solutions of non-linear PDE as an example. In a variational setting, a simple way to obtain solutions is by adding terms with (higher-order) derivatives, whose form is often dictated by modeling considerations. The prototypical example is

$$\int_\Omega W(u)dx + \varepsilon^2 \int_\Omega |\nabla u|^2 dx,$$

with W a non-convex potential. The limit of solutions u_ε to specific problems obtained using this energies are particular solutions to the corresponding problem with $\varepsilon = 0$ characterized by additional properties, the type of properties depending on the non-convexity of W, which can be of *double-well* type, or with a *non-convexity at infinity*.

Potentials of double-well type W are such that their convex envelope differs from W in bounded intervals, in which case, up to adding a linear term, we can consider W with two minimizers. This is the case of the Cahn–Hilliard theory of phase separations [14], for which a *minimal-interface criterion* was conjectured (e.g. by Gurtin [19]) stating that limit of minima satisfying (suitable) volume constraints take value only in the set of the two minimizers of W and the corresponding two sets are characterized as having a minimal interface of separation. In order to properly state such a convergence result one has to use the theory of *sets of finite perimeter* and of functions of bounded variation. The proof of the minimal-interface criterion was then achieved by Modica [22] using the Γ-convergence arguments by Modica and Mortola applied to energies (3), reinterpreting their limit as an interfacial energy, which can be written as

$$m_W \mathrm{Per}(E, \Omega);$$

that is, the *perimeter* of the set E in Ω, weighted by the *surface tension* m_W. The reason why the set E appears is that sequences u_ε with equibounded $F_\varepsilon(u_\varepsilon)$ are precompact in measure, so that by the boundedness of the first term in (3) their limit u satisfies $u(x) \in \{-1, 1\}$ and $u = -1 + 2\chi_E$, where $E = \{x : u(x) = 1\}$.

Potentials with a non-convexity at infinity have been treated in a different functional setting, considering perturbations of integrals depending on ∇u instead of depending on u. The simplest case is when W is the truncated quadratic potential $W(z) = \min\{|z|^2, 1\}$. After some scaling arguments, the perturbed functionals have the form

$$F_\varepsilon(u) = \frac{1}{\varepsilon} \int_\Omega W(\sqrt{\varepsilon}|\nabla u|)dx + \varepsilon^3 \int_\Omega |\nabla^2 u|^2 dx,$$

defined on $H^2(\Omega)$. The Γ-limit of these functionals has been studied by Alicandro et al. [2] and Bouchitté at al. [4] in the framework of the approximation of *free-discontinuity problems*. The Γ-limit is defined in the space of *special functions of bounded variation*, where it takes the form

$$F(u) = \int_\Omega |\nabla u|^2 dx + m_2^\infty \int_{S(u)\cap\Omega} \sqrt{|u^+(x) - u^-(x)|} d\mathcal{H}^{d-1},$$

where $S(u)$ is the set of *discontinuity points* of u and $u^\pm$ are the traces of u on both sides of $S(u)$. Note that if $d = 1$ then we can take $v = u'$ as a variable and reread F as a functional defined on spaces of measures, with domain measures μ which admit a decomposition as a sum of a part absolutely continuous with respect to the Lebesgue measure and a series of Dirac deltas. It is in order to avoid the description in spaces of measures that we use the variable u through its gradient.

Recent developments have made it possible to extend the theory of singular perturbations to the case with fractional derivatives of any order. In the case of double-well potentials, k integer and $s \in (0, 1)$, we study energies

$$\int_\Omega W(u)dx + \varepsilon^{2(k+s)} \int_{\Omega\times\Omega} \frac{|\nabla^k u(x) - \nabla^k u(y)|^2}{|x-y|^{d+2s}} dx\, dy, \tag{4}$$

where $\nabla^k u$ is the k-th order tensor of the partial derivatives of order k. In order to obtain a (local) perimeter functional, the correct scaling is obtained by dividing by ε, and considering the functionals

$$F_\varepsilon^{k+s}(u) = \frac{1}{\varepsilon} \int_\Omega W(u)dx + \varepsilon^{2(k+s)-1} \int_{\Omega\times\Omega} \frac{|\nabla^k u(x) - \nabla^k u(y)|^2}{|x-y|^{d+2s}} dx\, dy.$$

with $k + s > \frac{1}{2}$, so that the second term is a singular perturbation also after scaling. The limit is still a perimeter functional, which is in general of the form

$$F^{k+s}(E) = \int_{\partial^* E} \varphi_{k+s}(\nu) d\mathcal{H}^{d-1}, \tag{5}$$

where ν is the normal to E at the points in $\partial^* E$. In the one-dimensional case $d = 1$ the characterization of the perimeter functional reduces to that of the surface tension m_{k+s}. This paper examines the recent and less recent contributions by Alberti et al. [1], Fonseca and Mantegazza [16], Savin and Valdinoci [26], Palatucci and Vincini [24], Brusca et al. [11], Solci [28], Picerni [25], Braides et al. [8] in a unified setting, showing in particular how a common line of proof can be followed, that is also compatible with the extension to the integers case, formally when $s = 0$ and $k \geq 1$, and also to the limit degenerate case $s = \frac{1}{2}$.

The case of potentials with a non-convexity at infinity is much more complex since it requires the description of an interfacial energy φ also depending on the discontinuity size $|u^+(x) - u^-(x)|$. In a very complex work, using delicate localized interpolation inequalities Solci [27] has dealt with the higher-dimensional integer-derivative case

$$F_\varepsilon^k(u) = \frac{1}{\varepsilon}\int_\Omega W(u)dx + \varepsilon^{2k-1}\int_\Omega \|\nabla^k u\|^2 dx,$$

where $k \geq 2$ is an integer and $\|\cdot\|$ is the operator norm, showing that the limit is

$$F(u) = \int_\Omega |\nabla u|^2 dx + m_k^\infty \int_{S(u)} \sqrt[k]{|u^+(x) - u^-(x)|}\, d\mathcal{H}^{d-1},$$

with a proper characterization of m_k^∞. Note that the techniques introduced in [27] have been subsequently adapted to the double-well case in [11] and in [28], but the adaptation is not at all trivial. The case of fractional perturbations or norms different from the operator norm is almost completely open. In a final section of this paper we make a parallel with the results for the double-well case.

The content of this paper is a (very condensed) part of a course held in SISSA, Trieste in the 2024/2025 academic year in collaboration with M. Solci.

2 Singular Perturbations of Double-Well Energies

We focus on the one-dimensional case $d = 1$. The double-well potential $W : \mathbb{R} \to [0, +\infty)$ is a continuous function satisfying

(i) $W(z) = 0$ if and only if $z \in \{-1, 1\}$;
(ii) there exist the second derivatives $W''(-1)$ and $W''(1)$ and are strictly positive;
(iii) $\liminf\limits_{|z|\to+\infty} W(z) > 0$.

The scaled perturbed functionals take the form

$$\begin{aligned} F_\varepsilon^{k+s}(u) &= F_\varepsilon^{k+s}(u, I) \\ &= \frac{1}{\varepsilon}\int_I W(u)dx + \varepsilon^{2(k+s)-1}\int_{I\times I} \frac{|u^{(k)}(x) - u^{(k)}(y)|^2}{|x-y|^{1+2s}} dx\, dy, \end{aligned} \tag{6}$$

with domain the fractional Sobolev space $H^{k+s}(I)$ of the functions $u \in H^k(I)$ such that the double integral in (6) (the square of the corresponding Gagliardo seminorm of $u^{(k)}$) is finite; I is a bounded open interval of $\mathbb{R}$. We refer to [20] for an introduction to these fractional Sobolev spaces.

A key feature of energies (6) is their behaviour under scaling. More precisely, if we set $v(t) = u(\varepsilon t)$ then we have

$$F_\varepsilon^{k+s}(u) = F_1^{k+s}(v, \tfrac{1}{\varepsilon}I) = \int_{\frac{1}{\varepsilon}I} W(v)dx + \int_{\frac{1}{\varepsilon}I\times\frac{1}{\varepsilon}I} \frac{|v^{(k)}(x) - v^{(k)}(y)|^2}{|x-y|^{1+2s}} dx\, dy. \tag{7}$$

After this change of variables we can work on functionals depending on ε only through their domains. For convenience of notation, we write F^{k+s} in the place of F_1^{k+s}.

We will show that F_ε^{k+s} are equi-coercive with respect to the convergence in measure and their limit is a one-dimensional sharp-interface functional F^{k+s}, with domain piecewise-constant functions taking only the values -1 and 1, which we can identify with $BV(I; \{-1, 1\})$. On such functions, this functional is simply

$$F^{k+s}(u) = m_{k+s}\#(S(u)), \tag{8}$$

determined by the non-zero *surface tension* m_{k+s}. Here again $S(u)$ denotes the set of discontinuity points of u.

2.1 Coerciveness in Measure

For the classical Modica–Mortola result, coerciveness can be obtained by estimating the number of *transition intervals*; that is, with fixed $\eta \in (0, \frac{1}{2})$, intervals (x_1, x_2) such that $x_1 = -1 + \eta$ and $x_2 = 1 - \eta$. This is obtained by noting that the term $\int_{x_1}^{x_2} |u'|^2 dx$ is at least $\frac{1}{|x_2 - x_1|}$, so that the energy carried by such an interval is at least $C_\eta \frac{|x_2-x_1|}{\varepsilon} + \frac{\varepsilon}{|x_2-x_1|} \geq 2\sqrt{C_\eta} > 0$. This implies that $|x_2 - x_1| \sim \varepsilon$ and that the number of such transitions is equi-bounded. Applying the Bolzano–Weierstrass Theorem we can suppose that actually the transition intervals are independent of ε, and the compactness follows by noting that if $u_\varepsilon > 1 - \eta$ (or $u_\varepsilon < 1 - \eta$) on a fixed interval then u_ε tends to 1 (or to -1, respectively). Note that the optimization of this argument actually proves the lower bound with c_W.

In the general fractional case, if $k \geq 1$ this argument localized on transition intervals is not possible, as it fails simply considering functions u_ε affine on a transition interval, for which the Gagliardo seminorm is 0. The compactness argument must then use some non-local properties, which, loosely speaking, provide some extra boundary conditions to transition intervals, and is separate from the computation of the lower bound. It relies on the following local interpolation result by Solci [28, Lemma 6].

Lemma 1 (Local Interpolation Lemma) *If $F_\varepsilon^{k+s}(u_\varepsilon) \leq S < +\infty$ for all $\varepsilon > 0$, then there exists $\overline{\eta} > 0$ such that for all $\eta \in (0, \overline{\eta})$ there exists a constant L such that for all intervals J_ε such that $|J_\varepsilon| \geq \varepsilon L$ and $||u_\varepsilon| - 1| \leq \eta$ there exists at least a point $t \in J_\varepsilon$ such that $|u_\varepsilon^{(\ell)}(t)| \leq \varepsilon^{-\ell}\eta$ for all $\ell \in \{1, \dots, k-1\}$.*

Proof The proof follows by interpolation inequalities for fractional Sobolev spaces applied to either the function $u_\varepsilon - 1$ or $u_\varepsilon + 1$, after noting that by assumption (ii) on W there exist positive constants $\alpha, \beta, \overline{\eta}$ such that $\alpha(|z| - 1)^2 \leq W(z) \leq \beta(|z| - 1)^2$ if $||z| - 1| < \overline{\eta}$. □

We have the following compactness theorem [28, Theorem 2].

Theorem 1 (Compactness) *Let $\{u_\varepsilon\}$ be a family in $H^{k+s}(I)$ such that $F_\varepsilon(u_\varepsilon) \leq S < +\infty$ for all ε. Then, there exist $u \in BV(I;\{-1,1\})$ such that, up to subsequences, $u_\varepsilon \to u$ in measure. Furthermore, if W satisfies $W(z) \geq c_1|z|^p - c_2$ for some c_1, c_2 and $p \geq 1$, then $u_\varepsilon \to u$ in $L^p(I)$.*

Proof As already observed, it suffices to show that, for fixed η, u_ε has a finite number of transitions between $1-\eta$ and $-1+\eta$. Suppose otherwise, then for all ε small enough there exists an interval $J_\varepsilon = (a_\varepsilon, b_\varepsilon)$ containing a transition and such that $F_\varepsilon(u_\varepsilon, J_\varepsilon) = o(1)$ as $\varepsilon \to 0$, and $b_\varepsilon - a_\varepsilon > kL\varepsilon$. If J_ε contains up to k intervals such that $||u_\varepsilon| - 1| < \eta$ then we can apply Lemma 1. Since $b_\varepsilon - a_\varepsilon \leq C\varepsilon$ by the finiteness of the double-well term, we deduce that the functions $v_\varepsilon(t) = u_\varepsilon(a_\varepsilon + t(b_\varepsilon - a_\varepsilon))$ tend to a constant in $H^{s+k}(0,1)$, since they have vanishing H^{k+s} seminorm and $v_\varepsilon^{(\ell)}$ is equibounded at a point. This gives a contradiction with the transition boundary conditions.

If J_ε contains more such intervals then there exists an interval of length of order ε with a large number of transitions. Applying Rolle's theorem iteratively, we have the existence of points $t_1, \dots, t_\ell$ such that $u^{(\ell)}(t_\ell) = 0$ for all $\ell \in \{1, \dots, k-1\}$, and we can repeat the argument by contradiction.

Finally, if $W(z) \geq c_1|z|^p - c_2$ then $\int_{\{|u_\varepsilon|>2\}} |u_\varepsilon|^p dx \to 0$ and $(u_\varepsilon \vee -2) \wedge 2$ converges to u in $L^p(I)$, which proves that $u_\varepsilon \to u$ in $L^p(I)$. □

Remark 1 We note that in the proof of the compactness theorem instead of $-1+\eta$ and $1-\eta$ we can take any two numbers λ_1, λ_2 such that $[\lambda_1, \lambda_2] \cap \{-1, 1\} = \emptyset$. In particular, for every $r > 0$ we have that the number of transitions between $1+r$ and $1+2r$ is equibounded, and analogously for transitions between $-1-2r$ and $-1-r$.

2.2 *The Surface Tension*

The explicit characterization of the surface tension $m_1 = c_W$ (see (2)) in the Modica–Mortola result cannot be exported for other perturbations. It is convenient to characterize otherwise it in terms of an optimal-profile problem as in [17], where it is shown that

$$m_1 = \inf\Big\{\int_{\mathbb{R}} (W(v) + |v'|^2)dt : v(\pm\infty) = \pm 1\Big\}$$

$$= \inf_{T>0}\Big\{\int_{-T}^{T} (W(v) + |v'|^2)dt : v(\pm T) = \pm 1\Big\}. \tag{9}$$

Note that the second formula also gives an upper bound, suggesting the construction of recovery sequences of the form $u_\varepsilon(x) = v(\frac{x-\overline{x}}{\varepsilon})$ or $u_\varepsilon(x) = -v(\frac{x-\overline{x}}{\varepsilon})$ on intervals $(\overline{x} - \varepsilon T, \overline{x} + \varepsilon T)$ with $\overline{x} \in S(u)$ for $u \in BV(I;\{-1,1\})$.

This characterization of the surface tension can be exported to the fractional case as in the following proposition.

Proposition 1 **($k+s$-Surface Tension)** *Let $k \in \mathbb{N}$ and $s \in (0,1)$ with $k+s > \frac{1}{2}$. Then we have the equality*

$$m_{k+s} = \inf\Big\{ \int_{\mathbb{R}} W(v)dx + \int_{\mathbb{R}^2} \frac{|v^{(k)}(x) - v^{(k)}(y)|^2}{|x-y|^{1+2s}} dx\, dy : \\ v \in H^{k+s}_{\rm loc}(\mathbb{R}),\ \lim_{x\to\pm\infty} v(x) = \pm 1\Big\} \qquad (10)$$

$$= \inf_{T>0} \inf\Big\{ \int_{-T}^{+T} W(v)dt + \int_{\mathbb{R}^2} \frac{|v^{(k)}(x) - v^{(k)}(y)|^2}{|x-y|^{1+2s}} dx\, dy : \\ v \in H^{k+s}_{\rm loc}(\mathbb{R}), v(x) = \mathrm{sign}\, x \text{ if } |x| \ge T\Big\}. \qquad (11)$$

In order to prove this result and the lower bound for the Γ-limit, we preliminarily state a technical cut-off lemma, taken from the proof of [28, Proposition 12]. In the proof of the following lemma the cut-off functions φ are constructed in such a way that the resulting v, defined on intervals $(-T, T)$, are constant in a 'safe zone' of length L at the boundary of the interval. This is necessary in order to ensure that the additional contribution obtained by extending v to the whole $\mathbb{R}$ is small for L large.

Lemma 2 (A Cut-Off Lemma) *Let $\eta, \sigma, T, L > 0$, with $\eta < 1$, $u \in H^{k+s}(-T, T)$ such that $|u-1| \le \eta$ on $[\sigma, T]$ and $|u+1| \le \eta$ on $[-T, -\sigma T]$. Then there exists $v \in H^{k+s}_{\rm loc}(\mathbb{R})$ such that $v(x) = \mathrm{sign}(x)$ if $|x| \ge T - L$ and for all $\delta > 0$ there exists C_δ such that*

$$F^{k+s}(v, \mathbb{R})) \le (1+\delta)F^{k+s}(u, (-T, T)) \\ + \frac{C_\delta}{L^{2s}}\Big(F^{k+s}(u, (T-2L, T)) + F^{k+s}(v, (-T, -T+2L)) \\ + \frac{1}{\sigma T} + (1 + \sigma^{2s}T^{2s})F^{k+s}(u, (-T, T))\Big) + o(1) \qquad (12)$$

as $L \to +\infty$.

Proof The proof relies on a cut-off argument. The function v is defined by setting

$$v(x) = u(x)\,\varphi\big(\tfrac{|x|-T+2L}{L}\big) + \big(\mathrm{sign}(x) - \varphi\big(\tfrac{|x|-T+2L}{L}\big)\big),$$

where $\varphi \in C^\infty(\mathbb{R})$ with $0 \le \varphi \le 1$ is such that $\varphi(t) = 1$ if $t \le 0$, $\varphi(t) = 0$ if $t > 1$. The lemma follows after some lengthy computations, for which we refer to [28, Proposition 12]. The computations are particularly delicate since the estimate of $F^{k+s}(v, (-T, T))$ must be subdivided in many double integrals, for whose estimates different arguments must be used. An important point is to note that it is convenient to use that $|\varphi^{(\ell)}(x) - \varphi^{(\ell)}(y)| \le \frac{C}{L^\ell}|x-y|$ for $|x-y| \le 2L$

for $\ell \in \{0, \ldots, k-1\}$, while $|\varphi^{(\ell)}(x) - \varphi^{(\ell)}(y)| = 0$ if $|x - y| \geq 2L$ for $\ell \in \{1, \ldots, k-1\}$. □

Proof of Proposition 1 Since the expression in (10) gives a lower value than that in (11) as all test functions in (11) are also test functions for (10), it is sufficient to prove the converse inequality. To that end, with fixed $\eta > 0$, we consider a minimizer u for (10) up to η. We note that by Remark 1 applied to $u_\varepsilon(t) = u(\frac{t}{\varepsilon})$, the number of the transitions of u between $-1+\eta$ and $1-\eta$ is bounded, as well as those between $1+\eta$ and $1+2\eta$ and $-1-2\eta$ and $-1-\eta$. Moreover, the total measure of the set where $||u|-1| > \eta$ is finite. Hence, up to a translation, there exists x_0 such that $|u(x) - \mathrm{sign} x| < \eta$ for $|x| > x_0$ and $F^{k+s}(x_0, +\infty) + F^{k+s}(-\infty, -x_0) \leq \eta$. We fix $\sigma > 0$ and define $T = x_0/\sigma$. We can apply Lemma 2 with $L = T/3$, obtaining

$$\begin{aligned}F^{k+s}(v, \mathbb{R})) &\leq (1+\delta)F^{k+s}(u, (-T, T))\\ &\quad+\sigma^{2s}\frac{C_\delta}{x_0^{2s}}\Big(1 + \frac{1}{x_0} + (1 + x_0^{2s})F^{k+s}(u, (-T, T))\Big) + o(1),\end{aligned}$$

as $\sigma \to 0$, and the claim follows by the arbitrariness of σ and δ. □

2.3 The Γ-Limit

By the compactness theorem above, the functionals F_ε^{k+s} are equi-coercive with respect to the convergence in measure. Hence, the corresponding Γ-limit result can be stated on $BV(I; \{-1, 1\})$.

Theorem 2 (Sharp-Interface Limit) *Let $k \in \mathbb{N}$ and $s \in (0, 1)$ with $k + s > \frac{1}{2}$; then we have*

$$F^{k+s}(u) = m_{k+s}\#(S(u)). \tag{13}$$

for all $u \in BV(I; \{-1, 1\})$.

Proof Let $u_\varepsilon \to u$ in measure with $F_\varepsilon^{k+s}(u_\varepsilon) \leq S < +\infty$. Let $t_0 \in S(u)$, and define $\widetilde{u}_\varepsilon(t) = u_\varepsilon(t_0 + \varepsilon t)$. We have $F^{k+s}(\widetilde{u}_\varepsilon, (-T, T)) \leq S$ for all $T > 0$, and argue similarly to the proof of Proposition 1. Given $\eta > 0$, we fix x_0 such that, up to a change of sign, $|\widetilde{u}_\varepsilon(x) - \mathrm{sign}(x)| \leq \eta$ for $|x| \in [x_0, T]$. We can construct $v_\varepsilon \in H^{k+s}_{\mathrm{loc}}(\mathbb{R})$ such that $v_\varepsilon(x) = \mathrm{sign}(x)$ if $|x| \geq \frac{2}{3}T$ and for all $\delta > 0$ there exists C_δ such that

$$F^{k+s}(v_\varepsilon, \mathbb{R})) \leq (1+\delta)F_\varepsilon^{k+s}(u_\varepsilon, (t_0 - \varepsilon T, t_0 + \varepsilon T)) + \frac{C_\delta}{T^{2s}}.$$

Since v_ε is a test function for m_{k+s}, summing up for $t_0 \in S(u)$ and taking into account that all $(t_0 - \varepsilon T, t_0 + \varepsilon T)$ are disjoint for ε small enough, we obtain that

$$m_{k+s}\,\#(S(u)) \le (1+\delta)\liminf_{\varepsilon\to 0} F_\varepsilon^{k+s}(u_\varepsilon) + \frac{C_\delta}{T^{2s}}.$$

By the arbitrariness of T and δ we obtain the liminf inequality.

The construction of an approximate recovery sequence for a function u belonging to $BV(I;\{-1,1\})$ follows a usual pattern: given $\eta > 0$ we find $T > 0$ and $v \in H^{k+s}_{\rm loc}(\mathbb{R})$ such that $v(x) = \operatorname{sign} x$ if $|x| \ge T$ and

$$F^{k+s}(v,\mathbb{R}) \le m_{k+s} + \eta.$$

Let $S(u) = \{t_1, \dots, t_N\}$ and define

$$u_\varepsilon(t) = \begin{cases} v\Big(\frac{u(t_j^+)-u(t_j^-)}{2}\cdot\frac{t-t_j}{\varepsilon}\Big) & \text{if } |t-t_j|\le\varepsilon \\ u(t) & \text{otherwise.}\end{cases}$$

We set $x_0 = \inf I$, $x_j = (t_{j+1}+t_j)/2$ for $j \in \{1,\dots,N-1\}$ and $x_N = \sup I$. In order to check that u_ε is a recovery sequence, we need to show that the amount of energy due to the interactions between x and y belonging to intervals (x_{j_1-1}, x_{j_1}) and (x_{j_2-1}, x_{j_2}) with $j_1 \ne j_2$. If $k \ge 1$ we note that actually the interaction is only between the intervals $(t_{j_1} - \varepsilon T, t_{j_1} + \varepsilon T)$ and $(t_{j_2} - \varepsilon T, t_{j_2} + \varepsilon T)$, so that $|x-y| \sim |t_{j_2} - t_{j_1}|$ independently of ε, which gives that these contributions are asymptotically negligible. In the case $k = 1$ one additionally has to note that for $j_2 = j_1 + 1$ there is no interaction between pair of points in $(t_{j_1} + \varepsilon T, t_{j_2} - \varepsilon T)$. This implies that

$$F_\varepsilon^{k+s}(u_\varepsilon, I) \le (m_{k+s}+\eta)\#(S(u)) + o(1)$$

as $\varepsilon \to 0$, which proves the claim by the arbitrariness of η. □

2.4 Analysis at Critical Exponents

The results in the previous section show that the sharp-interface limit is described by the surface tension, which then defines a function $t \mapsto m_t := m_{k_t+s_t}$ for $t \in (\frac{1}{2}, +\infty)\setminus\mathbb{N}$, where $k_t = \lfloor t\rfloor$ and $s_t = t - \lfloor t\rfloor$ are the integer part and the fractional part of t, respectively. The behaviour of m_t at $\frac{1}{2}$ and at points in $\mathbb{N}$ is singular, in the sense that the limit of m_t is $+\infty$ at these points. In this section we briefly report results by Picerni [25] and Solci [28] showing how this singular behaviour can be 'corrected'.

2.4.1 Analysis Close the Exponent 1/2

If $k = 0$ and $s = 1/2$, we have $2(k + s) - 1 = 0$, so that the functional in (6)

$$\frac{1}{\varepsilon}\int_I W(u)dx + \int_{I\times I} \frac{|u(x)-u(y)|^2}{|x-y|^2}dx\,dy$$

presents a second term which is non-singular. Note that the second term is finite on $H^{1/2}(I)$, while the first term forces u to take the values -1 and 1, which is an incompatible requirement.

The second term in the functional above is scale invariant. As is often the case for scale-invariant energies, the correct scaling turns out to be logarithmic, and we define

$$F_\varepsilon^{1/2}(u) = \frac{1}{\varepsilon|\log\varepsilon|}\int_I W(u)dx + \frac{1}{|\log\varepsilon|}\int_{I\times I} \frac{|u(x)-u(y)|^2}{|x-y|^2}dx\,dy. \tag{14}$$

A classical result describing the asymptotic behaviour of $F_\varepsilon^{1/2}$ is the following (see [1, 26], to which we refer for a proof; see also [24]).

Theorem 3 *The functionals $F_\varepsilon^{1/2}$ are equi-coercive with respect to the convergence in measure, and the Γ-limit of $F_\varepsilon^{1/2}$ as $\varepsilon \to 0$ is the sharp-interface functional with domain $BV(I; \{-1, 1\})$*

$$F^{1/2}(u) = m_{1/2}\#(S(u))$$

with $m_{1/2} = 8$.

Despite having a different scaling, functionals $F_\varepsilon^{1/2}$ can be related to F_ε^s, using a more complex scaling factor than $\frac{1}{\varepsilon}$. This factor can be computed by examining the minimum problems for F_ε^s at varying ε and $s > 1/2$. The following result is proved in [25].

Theorem 4 (Correction at $s = \frac{1}{2}^+$) *The scaled functionals*

$$\begin{aligned}\widetilde{F}_\varepsilon^s(u) &:= \frac{2s-1}{1-\varepsilon^{2s-1}}F_\varepsilon^s(u)\\ &= \frac{2s-1}{\varepsilon(1-\varepsilon^{2s-1})}\Big(\int_I W(u(x))dx + \varepsilon^{2s}\int_I\int_I \frac{|u(x)-u(y)|^2}{|x-y|^{1+2s}}dx\,dy\Big)\end{aligned} \tag{15}$$

are equicoercive with respect to the convergence in measure, and they Γ-converge to $F^{1/2}$ for all $s = s_\varepsilon \to \frac{1}{2}^+$. In particular, we have that

$$\lim_{s\to 1/2^+}(2s-1)m_s = m_{1/2}.$$

Remark 2 (Interpolation Between Logarithmic and Linear Regimes) The factor $\frac{2s-1}{1-\varepsilon^{2s-1}}$ can be seen as an interpolation between the constant $2s-1$ (with $\varepsilon \to 0$ at s fixed) and the logarithmic scaling, since for fixed $\varepsilon > 0$ we have

$$\frac{2s-1}{1-\varepsilon^{2s-1}} = \frac{2s-1}{1-e^{-(2s-1)|\log\varepsilon|}} \sim \frac{1}{|\log\varepsilon|}$$

as $s \to \frac{1}{2}$. The asymptotic result can be interpreted as a Γ-expansion [9], which shows that

$$\int_I W(u(x))dx + \varepsilon^{2s}\int_I\int_I \frac{|u(x)-u(y)|^2}{|x-y|^{1+2s}}dx\,dy \sim \frac{8\varepsilon(1-\varepsilon^{2s-1})}{2s-1}\#(S(u))$$

on $BV(I;\{-1,1\})$ as $s \to \frac{1}{2}^+$ and $\varepsilon \to 0$.

2.4.2 Analysis at Positive Integers from the Left

A result by Bourgain, Brezis and Mironescu [5] shows that the functionals

$$(1-s)\int_{I\times I}\frac{|v(x)-v(y)|^2}{|x-y|^{1+2s}}dx\,dy$$

Γ-converge to $\int_I |v'|^2 dx$ as $s \to 1^-$. In particular if we do not multiply by the vanishing term $1-s$ the Γ-limit results finite only on constant functions, from which we obtain that m_s tends to $+\infty$ as $s \to 1^-$, showing a discontinuity with the Modica result. This discontinuity can be eliminated once we use the scaling by $1-s$, for which it can be shown (see [28]) that the limit of the *scaled surface tensions*

$$\widetilde{m}_s = \inf\Big\{\int_{\mathbb{R}} W(v)dx + (1-s)\int_{\mathbb{R}^2}\frac{|v(x)-v(y)|^2}{|x-y|^{1+2s}}dx\,dy : \\ v \in H^s_{\rm loc}(\mathbb{R}),\ \lim_{x\to\pm\infty} v(x) = \pm 1\Big\} \quad (16)$$

converge to m_1 as in (9) as $s \to 1^-$. In order to extend such a correction to all positive integers, we have to first make sure that a phase-transition limit exists also for (local) integer derivatives of order $k > 1$. This has been proved by Fonseca and Mantegazza [16] for second derivatives and only recently by Brusca, Donati and Solci [11] in the general case.

Theorem 5 (Phase Transitions with Integer-Order Perturbations) *Let W satisfy the same hypotheses as in the results already stated, and let $k \geq 1$ be a positive* *integer. Then the functionals*

$$F_\varepsilon^k(u) = \frac{1}{\varepsilon}\int_I W(u)dx + \varepsilon^{2k-1}\int_I |u^{(k)}(x)|^2 dx, \tag{17}$$

with domain $H^k(I)$ are equicorcive in measure, and their Γ-limit with respect both to the convergence on measure and to the L^2-convergence is the sharp-interface functional $m_k\#(S(u))$ with domain $BV(I;\{-1,1\})$, where

$$\begin{aligned} m_k &= \inf\Big\{\int_{\mathbb{R}} W(v)dx + \int_{\mathbb{R}} |v^{(k)}|^2 dx : v\in H^k_{\rm loc}(\mathbb{R}),\ \lim_{x\to\pm\infty} v(x) = \pm 1\Big\} \\ &= \inf_{T>0}\inf\Big\{\int_{-T}^{+T} W(v)dt + \int_{-T}^{T} |v^{(k)}(x)|^2 dx\ : v\in H^k(-T,T), \\ &\qquad v(\pm T) = \pm 1 \text{ and } v^{(\ell)}(x) = 0 \text{ if } |x| = T, \ell\in\{1,\dots,k-1\}\Big\}. \end{aligned} \tag{18}$$

Proof It suffices to note that the line of proof for the energies F_ε^{k+s} works almost unchanged, with actually some simplifications in the computations when using a cut-off argument. □

Remark 3 (Enhanced Boundary Conditions) We note that, contrary to (11), the second formula in (18) uses minimum problems on the bounded intervals $(-T, T)$ with homogeneous boundary conditions on the derivatives. Contrary to the case $k = 1$, this has to be achieved using local interpolation inequalities as in Lemma 1 [11, 28].

In order to correct the behaviour of surface tensions at integer points 'from the left' we define the Bourgain-Brezis-Mironescu scaled surface tensions

$$m_{k+s}^{BBM} = \inf\Big\{\int_{\mathbb{R}} W(v)dx + (1-s)\int_{\mathbb{R}^2}\frac{|v^{(k)}(x)-v^{(k)}(y)|^2}{|x-y|^{1+2s}}dx\,dy : \\ v\in H^{k+s}_{\rm loc}(\mathbb{R}),\ \lim_{x\to\pm\infty} v(x) = \pm 1\Big\}. \tag{19}$$

Note that for s fixed the surface tension describes the Γ-limit for the corresponding scaled functionals. To check this, it suffices to note that $(1-s)^{-1}W$ is a potential satisfying the same assumptions as W. We then have the following result.

Proposition 2 (Continuity from the Left) *For all $k \geq 0$ non-negative integers we have* $\lim_{s\to 1^-} m_{k-1+s}^{BBM} = m_k$.

Proof The proof relies on the Bourgain, Brezis and Mironescu result applied with $u^{(k-1)}$ in the place of v, and on the use of Lemma 2, which guarantees that, up to a uniformly small error, minimum problems for m_{k-1+s}^{BBM} can be taken with a fixed large T independent of s. □

2.4.3 Analysis at Positive Integers from the Right

The result by Bourgain, Brezis and Mironescu has a counterpart for $s \to 0^+$ proved by Maz'ya and Shaposhnikova [21]. In this case the functionals

$$s \int_{\mathbb{I}\times\mathbb{I}} \frac{|v(x)-v(y)|^2}{|x-y|^{1+2s}} dx\, dy \tag{20}$$

Γ-converge to $2\int_I |v|^2 dx$ as $s \to 0^+$ on functions u with compact support in $\mathbb{I}$. This restriction on the support is not relevant for our analysis since, by the restriction $k+s > \frac{1}{2}$, the result is applied to functions $v = u^{(k)}$ with $k \geq 1$, with the condition that u is constant (either -1 or 1) outside $\mathbb{I}$, and hence v is can be assumed with compact support in $\mathbb{I}$. Again, if we do not multiply by the vanishing term s, the Γ-limit results finite only on constant functions, from which we obtain that m_{k+s} tends to $+\infty$ as $s \to 0^+$, showing a discontinuity with the result by Brusca, Donati, and Solci. This discontinuity can be eliminated once we use the scaling by s in the following way.

For $k \geq 1$ and $s \in (0,1)$, we define the Maz'ya and Shaposhnikova scaled surface tensions

$$m^{MS}_{k+s} = \inf\Big\{ \int_{\mathbb{R}} W(v)dx + \frac{s}{2}\int_{\mathbb{R}^2} \frac{|v^{(k)}(x)-v^{(k)}(y)|^2}{|x-y|^{1+2s}} dx\, dy : \\ v \in H^{k+s}_{\rm loc}(\mathbb{R}),\ \lim_{x\to\pm\infty} v(x) = \pm 1\Big\}. \tag{21}$$

The normalization coefficient 2 is due to the factor appearing in the Γ-limit of (20) as $s \to 0^+$. Again note that for s fixed the surface tension describes the Γ-limit for the corresponding scaled functionals, since $2s^{-1}W$ is a potential satisfying the same assumptions as W. We then have the following result.

Proposition 3 (Continuity from the Right) *For all non-negative integers $k \geq 1$ we have* $\lim_{s\to 0^+} m^{MS}_{k+s} = m_k$.

Proof The proof is the same as for $s \to 1^-$ but using the Maz'ya and Shaposhnikova result applied with $u^{(k)}$ in the place of v. □

Remark 4 (Continuous Interpolations for $k+s > \frac{1}{2}$) Taking into account the 'corrections' for $k \geq 1$ and $s \to 0^+$ and $s \to 1^-$ at the same time, one can construct functionals

$$\int_I W(u)dx + \varepsilon^{2(k+s)}\alpha_s \int_{\mathbb{R}^2} \frac{|v^{(k)}(x)-v^{(k)}(y)|^2}{|x-y|^{1+2s}} dx\, dy$$

that interpolate uniformly between functionals (17) in the sense of expansion by Γ-convergence [9]. In [28] the choice is $\alpha_s = s(1-s)2^{s-1}$.

2.5 Some Notes on the Higher-Dimensional Case

The higher-dimensional case, when we replace the interval I with an open set Ω in $\mathbb{R}^d$ and the functionals have the form

$$F_\varepsilon^{k+s}(u) = \frac{1}{\varepsilon}\int_\Omega W(u)dx + \varepsilon^{2(k+s)-1}\int_{\Omega\times\Omega}\frac{\|\nabla^k u(x) - \nabla^k u(y)\|^2}{|x-y|^{d+2s}}dx\,dy, \tag{22}$$

is more technical. We note however that the compactness result follows from the one-dimensional result by Solci.

Proposition 4 (Equi-Coerciveness) *Let $k \geq 0$, $s \in (0,1)$ with $k+s > \frac{1}{2}$, and let u_ε be such that $F_\varepsilon^{k+s}(u_\varepsilon) \leq S < +\infty$. We assume that the potential W also satisfies a 2-polynomial growth condition. Then, up to subsequences, u_ε converges in $L^2(\Omega)$ to some $u \in BV(\Omega; \{-1, 1\})$.*

Proof The proof follows from the one-dimensional result by a characterization by slicing of higher-order fractional Sobolev spaces [20]. We refer to [8] for details. □

Identifying a function $u \in BV(\Omega; \{-1, 1\})$ with the set $E = \{u = 1\}$ such that $u = -1 + 2\chi_E$, the Γ-limit of F_ε^{k+s} can be written as a perimeter functional

$$F_{k+s}(E) = \int_{\partial^* E}\varphi_{k+s}(\nu_E)d\mathcal{H}^{d-1},$$

with $\partial^* E$ the reduced boundary of E and ν_E the internal normal to E. The integrand φ_{k+s} is a function such that its positively homogeneous extension of degree 1 is convex (see [3]), and can be characterized through some asymptotic formulas [8, 15]. The form of φ_{k+s} depends on the interplay between the norm $\|\cdot\|$ of a k-th order tensor and the non-locality of the Gagliardo seminorm, except for the case $k = 0$, treated in [26], since in that case we have the norm of a 0-th order tensor; that is, of a scalar quantity.

With respect to the one-dimensional case, the comparison with the local perturbation by the k-th derivative seems more complex. That local case is part of the work of Brusca, Donati and Solci [11], when $\|\cdot\|$ is the operator norm and an argument by sections can be applied, and has been generalized to arbitrary norms by Brusca, Donati and Trifone [12]. In particular, the 'correction' of the non-local energies at integer points seems to require new arguments, at least when the limit energy density φ_k derived in the local case is not a constant.

3 Singular Perturbations of Energies with Non-convexity at Infinity

The second type of non-convex potentials W are those with a non-convexity at infinity; that is, such that the convex envelope of W does not coincide with W on one or two half lines. Typical potentials with a well at infinity are the Lennard-Jones potential and the Perona-Malik potential $W(z) = \log(1+z^2)$. For such potentials the results are still partial and sparse. In this section we analyze the few known results trying to draw some parallel with the previous sections. We will mainly consider a one-dimensional environment.

3.1 Perturbations of Local Type

Up to a translation, it is not restrictive to suppose that W has a minimum point in 0 and that $W(0) = 0$. A prototypical W is the *truncated quadratic potential* $W(z) = \min\{z^2, 1\}$. Since most of the issues are open and independent of the shape of the potential, we will restrict our analysis to this particular one.

As remarked in the Introduction, it is convenient to use u' as a variable in the place of u, and suitably scale the energies, in order to avoid the necessity of defining the Γ-limit on spaces of measures. We begin with the perturbation of minimal (integer) order; that is, with a second derivative. After scaling, the energies we consider have the form

$$\int_I \frac{1}{\varepsilon} W(\sqrt{\varepsilon} u')dx + \varepsilon^3 \int_I |u''|^2 dx. \tag{23}$$

The power 3 is in accord with the scaling arguments already used for higher-order derivatives, as $3 = 2k - 1$, with $k = 2$. The reason why for the first integral we consider the scaling in (23) is that, assuming that $W''(0) > 0$, we have

$$\frac{1}{\varepsilon} W(\sqrt{\varepsilon} u') = \frac{1}{\varepsilon}\Big(\frac{1}{2} W''(0)\varepsilon |u'|^2 + o(\varepsilon |u'|^2)\Big) = \frac{1}{2} W''(0)|u'|^2 + o(|u'|^2),$$

which suggests in a way that $u' \in L^2(I)$. Note that this scaling is not the unique choice (see e.g. [18]), but in a sense is the most complete and justified as an expansion by Γ-convergence [9]. For the prototypical quadratic energies

$$F^2_\varepsilon(u) = \int_I \min\Big\{|u'|^2, \frac{1}{\varepsilon}\Big\} dx + \varepsilon^3 \int_I |u''|^2 dx \tag{24}$$

indeed, it turns out that sequences with equi-bounded energy are precompact with respect to the L^1-convergence (up to addition of constants), and the Γ-limit of F^2_ε

in this topology is

$$F_2(u) = \int_I |u'|^2 dx + m_2 \sum_{S(u)} \sqrt{|u^+(x) - u^-(x)|}$$

defined in the space $SBV(I)$ (see [7]). This has been proved by Alicandro et al. [2] and by Bouchitté et al. [4].

The constant m_2 represents the energy necessary to produce a jump discontinuity of size 1. We reduce to this case by a $1/2$-homogeneity argument. It is important to note that m_2 can be interpreted as given by an optimal-profile problem

$$m_2 = \inf_T \min \left\{ T + \int_0^T |v''|^2 dt : v(0) = 0, v(T) = 1, v'(0) = v'(T) = 0 \right\},$$

where the transition between 0 and 1 is optimized on an interval of length T, which is also a variable of the problem. This minimum problem is obtained after scaling by $1/\varepsilon$ an interval where the modulus of the derivative of a function u_ε is *above the threshold* $1/\sqrt{\varepsilon}$. After defining $v(t) = u(\varepsilon t)$, we obtain the homogeneous boundary conditions on v' at the endpoints as $\varepsilon \to 0$. This consideration draws a parallel with the phase-transition energies, in which we have observed that boundary conditions in (9) are automatically determined in the Modica–Mortola case, while this is not the case for higher-order perturbations.

The general result for perturbations with local higher-order energies is the following theorem by Solci [27].

Theorem 6 (Integer Higher-Order Perturbations) *Let $k \geq 2$, $k \in \mathbb{N}$, and let*

$$F^k_\varepsilon(u) = \int_I \min\left\{ |u'|^2, \frac{1}{\varepsilon} \right\} dx + \varepsilon^{2k-1} \int_I |u^{(k)}|^2 dx \qquad (25)$$

be defined on $H^k(I)$. Then the functionals are equi-coercive, up to addition of constants, with respect to the convergence in $L^1(I)$, the domain of the limit is a subset of $SBV(I)$, where it takes the form

$$F_k(u) = \int_\Omega |\nabla u|^2 dx + m_k \sum_{S(u)} \sqrt[k]{|u^+(x) - u^-(x)|},$$

and the constant m_k is defined by

$$m_k = \inf_T \min \left\{ T + \int_0^T |v^{(k)}|^2 dt : v(0) = 0, v(T) = 1, \right.$$
$$\left. v^{(\ell)}(0) = v^{(\ell)}(T) = 0 \; for \; \ell \in \{1, \ldots, k-1\} \right\}. \qquad (26)$$

The proof of this result is extremely delicate, and we refer to [28] for the details of the proof. Again, one issue is, as for phase-transition energies, to enhance the boundary conditions, showing that one can add also conditions on the derivatives of order higher than the first. This can be done using interpolation inequalities, possibly enlarging the intervals where functions have to be considered 'above the threshold'. Unfortunately, the length of these intervals itself concurs in the determination of m_k, and the corresponding minimum problems can be extremely sensitive to the change of length if the jump size is small, when the $1/k$-homogeneity argument hinted at above is applied. This second issue can be solved by fine constructions when we have small jumps, and by the use of relaxation results.

Remark 5 (The Higher-Dimensional Case) In [27] the higher-dimensional case has been treated for a perturbation using the operator norm of the tensor of the k-th order derivatives, obtaining F_k with the corresponding form as in (25). The case of norms different from the operator one is open. Note in particular that it is not clear how to generalize the formula for m_k to obtain a function φ_k, which now may depend not only on the normal to $S(u)$ but also to the jump size. We note that the result with the operator norm is anyhow sufficient to obtain equi-coerciveness in the general case by comparison.

3.2 A Result for Fractional Perturbations

While the case of general higher-order fractional perturbations seems technically complex, we can treat the case when $k = 1$ and $s \in (0, 1)$, which may be regarded as a variation of the result in [2], and does not require an extension to the fractional case of the interpolation techniques of [27]. The functionals

$$F_\varepsilon^{1+s}(u) = \int_I \min\Big\{|u'|^2, \frac{1}{\varepsilon}\Big\}dx + \varepsilon^{2s+1}\int_I\int_I \frac{|u'(x) - u'(y)|^2}{|x - y|^{1+2s}}dx\,dy,$$

defined on $H^{1+s}(I)$, have been studied by Braides and Vitali [10]. Note that the exponent $2s + 1 = 2(k + s) - 1$ with $k = 1$ is determined by the usual scaling argument.

With (26) in mind, the corresponding constant m_{1+s} can be written in the form of (11) with the non-local part integrated on the whole line, as

$$m_{1+s} = \inf_T \min\Big\{T + \int_{\mathbb{R}}\int_{\mathbb{R}} \frac{|v'(x) - v'(y)|^2}{|x - y|^{1+2s}}dx\,dy : v \in H^{1+s}_{\rm loc}(\mathbb{R})$$
$$v(t) = 0 \text{ for } t \le 0,\ v(t) = 1 \text{ for } t \ge T\Big\}.$$

With this definition, we can describe the asymptotics of F_ε^{1+s} as $\varepsilon \to 0$ as follows.

Theorem 7 *The functionals F_ε^{1+s} are equi-coercive, up to addition of constants, with respect to the $L^1(I)$-convergence. The corresponding Γ-limit is*

$$F_{1+s}(u) = \int_I |u'|^2 dx + m_{1+s} \int_{S(u)} |u^+(x) - u^-(x)|^{\frac{1}{1+s}} d\mathcal{H}^{d-1},$$

defined on $SBV(I)$.

For a proof we refer to [10]. We only note that if $s > \frac{1}{2}$ the boundary conditions $v'(0) = v'(T) = 0$ are (meaningful and) enough to obtain compactness following the argument of [2], but limiting the double integral to $(0, T) \times (0, T)$ in the definition of m_{1+s} gives a sub-optimal lower bound.

Remark 6 (Some Open Problems) Beside the extension to higher-order fractional derivatives and to higher dimension, we note that it could be interesting to analyze the behaviour at integer points, in particular starting from the one-dimensional case with $k = 1$ and s tending to 0 or tending to 1. Note that in the first case we expect a limit functional of linear growth defined on $BV(I)$.

Acknowledgments The author gratefully acknowledges the active participation of the students of the course at SISSA. The author is a member of GNAMPA of INdAM.

References

1. Alberti, G., Bouchitté, G., Seppecher, P.: Un résultat de perturbations singulières avec la norme $H^{1/2}$. C. R. Acad. Sci. Paris Sér. I Math. **319**(4), 333–338 (1994)
2. Alicandro, R., Braides, A., Gelli, M.S.: Free-discontinuity problems generated by singular perturbation. Proc. Roy. Soc. Edinburgh Sect. A **128**(6), 1115–1129 (1998)
3. Ambrosio, L., Braides, A.: Functionals defined on partitions in sets of finite perimeter. II. Semicontinuity, relaxation and homogenization. J. Math. Pures Appl. (9) **69**(3), 307–333 (1990)
4. Bouchitté, G., Dubs, C., Seppecher, P.: Regular approximation of free-discontinuity problems. Math. Models Methods Appl. Sci. **10**(07), 1073–1097 (2000)
5. Bourgain, J., Brezis, H., Mironescu, P.: Another look at Sobolev spaces. In: Optimal Control and Partial Differential Equations, pp. 439–455. IOS, Amsterdam (2001)
6. Braides, A.: Homogenization of noncoercive integrals. Ricerche Mat. **32**(2), 347–368 (1983)
7. Braides, A.: Approximation of Free-discontinuity Problems. Springer-Verlag, Berlin (1998)
8. Braides, A., Oleinik, R., Picerni, M.: Higher-order non-local gradient theory of phase-transitions: the higher-dimensional case. In preparation
9. Braides, A., Truskinovsky, L.: Asymptotic expansions by Γ-convergence. Contin. Mech. Thermodyn. **20**(1), 21–62 (2008)
10. Braides, A., Vitali, E.: Fractional approximation of free-discontinuity problems. In preparation
11. Brusca, G.C., Donati, D., Solci, M.: Higher-order singular perturbation models for phase transitions. SIAM J. Math. Anal. **57**(3), 3146–3170 (2025)
12. Brusca, G.C., Donati, D., Trifone, C.: Singular perturbations models in phase transitions for anisotropic higher-order materials. Calc. Var. Partial Differ. Equ. **64**(243), (2025)

13. Buttazzo, G., Dal Maso, G.: Γ-limit of a sequence of nonconvex and non-equi-Lipschitz integral functionals. Ricerche Mat. **27**(2), 235–251 (1978)
14. Cahn, J.W., Hilliard, J.E.: Free energy of a nonuniform system. I. Interfacial free energy. J. Chem. Phys. **28**(2), 258–267 (1958)
15. Dal Maso, G., Fonseca, I., Leoni, G.: Asymptotic analysis of second order nonlocal Cahn–Hilliard-type functionals. Trans. Am. Math. Soc. **370**(4), 2785–2823 (2018)
16. Fonseca, I., Mantegazza, C.: Second order singular perturbation models for phase transitions. SIAM J. Math. Anal. **31**(5), 1121–1143 (2000)
17. Fonseca, I., Tartar, L.: The gradient theory of phase transitions for systems with two potential wells. Proc. Roy. Soc. Edinburgh Sect. A **111**(1–2), 89–102 (1989)
18. Gobbino, M., Picenni, N.: A quantitative variational analysis of the staircasing phenomenon for a second order regularization of the Perona-Malik functional. Trans. Am. Math. Soc. **376**(8), 5307–5375 (2023)
19. Gurtin, M.E.: Some results and conjectures in the gradient theory of phase transitions. In: Metastability and Incompletely Posed Problems, pp. 135–146. Springer, Berlin (1987)
20. Leoni, G.: A First Course in Fractional Sobolev Spaces, vol. 229. Graduate Studies in Mathematics. American Mathematical Society, Providence (2023)
21. Maz'ya, V., Shaposhnikova, T.: On the Bourgain, Brezis, and Mironescu theorem concerning limiting embeddings of fractional Sobolev spaces. J. Funct. Anal. **195**(2), 230–238 (2002)
22. Modica, L.: The gradient theory of phase transitions and the minimal interface criterion. Arch. Ration. Mech. Anal. **98**, 123–142 (1987)
23. Modica, L., Mortola, S.: Un esempio di Γ^--convergenza. Boll. Un. Mat. Ital. B (5) **14**(1), 285–299 (1977)
24. Palatucci, G., Vincini, S.: Gamma-convergence for one-dimensional nonlocal phase transition energies. Matematiche (Catania) **75**(1), 195–220 (2020)
25. Picerni, M.: Analysis for non-local phase transitions close to the critical exponent $s = \frac{1}{2}$. Ric. Mat. **74**(3), 1599–1626 (2025)
26. Savin, O., Valdinoci, E.: Γ-convergence for nonlocal phase transitions. Ann. Inst. H. Poincaré C Anal. Non Linéaire **29**(4), 479–500 (2012)
27. Solci, M.: Local interpolation techniques for higher-order singular perturbations of non-convex functionals: Free-discontinuity problems. J. Math. Pures Appl. (9) **204**, Paper No. 103776 (2025)
28. Solci, M.: Higher-order non-local gradient theory of phase-transitions. Milan J. Math. **93**(2), 455–486 (2025)

A Survey on the Div-Curl Lemma and Some Extensions to Fractional Sobolev Spaces

Maicol Caponi

Abstract This chapter recalls the classical formulation of the Div-Curl lemma along with its proof, and presents some possible generalizations in the fractional setting, within the framework of the Riesz fractional gradient and divergence introduced by Shieh and Spector (Adv Calc Var 8:321–336;2015) and further developed by Comi and Stefani (J Funct Anal 277:3373–3435;2019).

1 Introduction

Partial differential equations arising in continuum mechanics are often nonlinear, which creates significant analytical challenges. One of the main difficulties is that weak convergence, commonly used in variational problems, does not behave well with respect to nonlinear operations. In particular, the product of two weakly converging sequences does not necessarily converge to the product of their limits.

The theory of *Compensated Compactness*, developed by Murat and Tartar in the late 1970s (see [16, 17, 22–24]), provides a powerful framework to overcome this problem. Its starting point, the *Div-Curl lemma*, shows that the scalar product of two weakly converging sequences can itself converge weakly, under suitable conditions on their divergence and curl. More generally, the key idea of *Compensated Compactness* is that certain nonlinear expressions may exhibit weak continuity when appropriate differential constraints are imposed.

Originally introduced in the context of *homogenization*, the *Div-Curl lemma* was first used to prove compactness results for nonlinear differential operators with oscillating coefficients. It soon found applications in other areas, such as *conservation laws*, where it helps to derive entropy solutions via Young measures,

M. Caponi (✉)
Dipartimento di Ingegneria e Scienze dell'Informazione e Matematica, Università degli Studi dell'Aquila, L'Aquila, Italy
e-mail: maicol.caponi@univaq.it

J. Duran i Lamiel, A. Maione (eds.), *New Frontiers in Homogenization and Fractional Calculus*, Trends in Mathematics,
https://doi.org/10.1007/978-3-032-20098-3_2

and *nonlinear elasticity*, where it can be used to show the weak continuity of the determinant of the Jacobian matrix.

The aim of this chapter is twofold. First, in Sect. 2 we recall the classical formulation of the Div-Curl lemma and present its proof, following the approach in [10] (see also [16]), which is based on Rellich's compactness theorem and standard local elliptic regularity estimates. Second, in Sect. 3 we discuss several extensions of the Div-Curl lemma to the fractional setting of [6, 18]. These nonlocal versions may be useful for analyzing the compactness properties of fractional differential operators with oscillating coefficients, and could find applications in the study of nonlocal partial differential equations.

2 The Div-Curl Lemma

In this section, we introduce the notation that will be used throughout the chapter. We then recall the classical formulation of the Div-Curl lemma and provide a complete proof, based on Rellich's compactness theorem and standard elliptic regularity estimates. This argument will serve as a reference point for the nonlocal generalizations discussed in the next section.

Let $n \geq 2$, let $\Omega \subset \mathbb{R}^n$ be an open set, and let $p \in (1, \infty)$ be fixed. We denote by $p' := \frac{p}{p-1} \in (1, \infty)$ the Hölder conjugate exponent of p. Given two open sets $A, B \subset \mathbb{R}^n$, we write $A \Subset B$ if there exists a compact set $K \subset \mathbb{R}^n$ such that $A \subset K \subset B$. The space of distributions on Ω, namely the continuous dual space of $C_c^\infty(\Omega)$ endowed with the strong dual topology, is denoted by $\mathcal{D}'(\Omega)$.

The Lebesgue and Sobolev spaces are defined as usual. In particular, we consider

$$W_0^{1,p}(\Omega) := \overline{C_c^\infty(\Omega)}^{\|\cdot\|_{W^{1,p}(\mathbb{R}^n)}}, \qquad W^{-1,p}(\Omega) := (W_0^{1,p'}(\Omega))',$$

as well as the local Sobolev spaces

$$W_{\mathrm{loc}}^{1,p}(\Omega) := \{u \in L_{\mathrm{loc}}^p(\Omega) \,:\, \nabla u \in L_{\mathrm{loc}}^p(\Omega; \mathbb{R}^n)\},$$
$$W_{\mathrm{loc}}^{-1,p}(\Omega) := \{f \in \mathcal{D}'(\Omega) \,:\, f \in W^{-1,p}(\Omega') \text{ for all open sets } \Omega' \Subset \Omega\}.$$

Remark 1 Given a function $u \in L_{\mathrm{loc}}^p(\Omega)$, then both u and its distributional derivatives $\partial_i u$ (in the sense of distributions) belong to $W_{\mathrm{loc}}^{-1,p}(\Omega)$ for all $i \in \{1, \ldots, n\}$. In particular, for all bounded open sets $\Omega' \Subset \Omega$, we have

$$\langle u, v \rangle_{W^{-1,p}(\Omega') \times W_0^{1,p'}(\Omega')} = \int_{\Omega'} uv \,\mathrm{d}x \qquad \text{for all } v \in W_0^{1,p'}(\Omega'),$$

$$\langle \partial_i u, v \rangle_{W^{-1,p}(\Omega') \times W_0^{1,p'}(\Omega')} = -\int_{\Omega'} u \partial_i v \,\mathrm{d}x \qquad \text{for all } v \in W_0^{1,p'}(\Omega').$$

These identities imply the following estimates

$$\|u\|_{W^{-1,p}(\Omega')} \leq \|u\|_{L^p(\Omega')}, \qquad \|\partial_i u\|_{W^{-1,p}(\Omega')} \leq \|u\|_{L^p(\Omega')}.$$

Definition 1 Given a vector field $F \in L^p_{\mathrm{loc}}(\Omega; \mathbb{R}^n)$, we define $\operatorname{div} F \in W^{-1,p}_{\mathrm{loc}}(\Omega)$ and $\operatorname{curl} F \in W^{-1,p}_{\mathrm{loc}}(\Omega, \mathbb{R}^{n\times n})$, respectively, by

$$\operatorname{div} F := \sum_{j=1}^{n} \partial_j F_j, \qquad (\operatorname{curl} F)_{ij} := \partial_j F_i - \partial_i F_j \quad \text{for all } i, j \in \{1, \dots, n\}.$$

Similarly, given a matrix field $\mathbf{A} \in L^p_{\mathrm{loc}}(\Omega; \mathbb{R}^{n\times n})$, we define $\operatorname{div} \mathbf{A} \in W^{-1,p}_{\mathrm{loc}}(\Omega; \mathbb{R}^n)$ by

$$(\operatorname{div} \mathbf{A})_i := \sum_{j=1}^{n} \partial_j \mathbf{A}_{ij} \quad \text{for all } i \in \{1, \dots, n\}.$$

We are now ready to state the *Div-Curl lemma*.

Theorem 1 (Div-Curl Lemma) *Let* $(F^k)_k \subset L^p_{\mathrm{loc}}(\Omega; \mathbb{R}^n)$, $F^\infty \in L^p_{\mathrm{loc}}(\Omega; \mathbb{R}^n)$, $(G^k)_k \subset L^{p'}_{\mathrm{loc}}(\Omega; \mathbb{R}^n)$, *and* $G^\infty \in L^{p'}_{\mathrm{loc}}(\Omega; \mathbb{R}^n)$ *satisfy as* $k \to \infty$

$$F^k \to F^\infty \quad \textit{weakly in } L^p_{\mathrm{loc}}(\Omega; \mathbb{R}^n), \qquad G^k \to G^\infty \quad \textit{weakly in } L^{p'}_{\mathrm{loc}}(\Omega; \mathbb{R}^n). \tag{1}$$

Assume that as $k \to \infty$

$$\operatorname{div} F^k \to \operatorname{div} F^\infty \qquad \textit{strongly in } W^{-1,p}_{\mathrm{loc}}(\Omega), \tag{2}$$

$$\operatorname{curl} G^k \to \operatorname{curl} G^\infty \qquad \textit{strongly in } W^{-1,p'}_{\mathrm{loc}}(\Omega; \mathbb{R}^{n\times n}). \tag{3}$$

Then, as $k \to \infty$

$$F^k \cdot G^k \to F^\infty \cdot G^\infty \quad \textit{weakly* in } \mathcal{D}'(\Omega). \tag{4}$$

Remark 2 Let us briefly comment on Theorem 1.

(a) The Div-Curl lemma is sometimes stated by requiring that the sequences $(\operatorname{div} F^k)_k$ and $(\operatorname{curl} G^k)_k$ are uniformly bounded in $L^p_{\mathrm{loc}}(\Omega)$ and $L^{p'}_{\mathrm{loc}}(\Omega; \mathbb{R}^{n\times n})$, respectively. Due to the compact embedding $L^p_{\mathrm{loc}}(\Omega) \subset W^{-1,p}_{\mathrm{loc}}(\Omega)$, these uniform bounds imply the strong convergence assumptions (2) and (3). In this case, the Div-Curl lemma can be interpreted as follows: if two sequences converge weakly, and if a suitable combination of their derivatives (namely, their divergence and curl) remain bounded, then the nonlinear expression given by their scalar product converges, in the sense of distributions, to the product of the weak limits.

(b) In the case $p = 2$, Theorem 1 can be proved in a simple way by using the Fourier transform (this was, indeed, the original approach of Murat and Tartar). The interested reader can find this argument, for instance, in [25, Lemma 7.2].

(c) Under the assumptions of Theorem 1, we deduce that the sequence $(F^k \cdot G^k)_k$ is uniformly bounded in $L^1_{\mathrm{loc}}(\Omega)$. In particular, $(F^k \cdot G^k)_k$ is uniformly bounded in the space of Radon measures $\mathcal{M}(\Omega) := (C_c(\Omega))'$. Therefore, by the Banach-Alaoglu's theorem and by exploiting the Div-Curl lemma to identify the weak* limit in $\mathcal{M}(\Omega)$, we conclude that, as $k \to \infty$

$$F^k \cdot G^k \to F^\infty \cdot G^\infty \quad \text{weakly* in } \mathcal{M}(\Omega).$$

One may hope to improve this convergence, for instance to have as $k \to \infty$

$$F^k \cdot G^k \to F^\infty \cdot G^\infty \quad \text{weakly in } L^1_{\mathrm{loc}}(\Omega).$$

However, this is generally false. Indeed, as shown in [25, Lemma 7.3], one can construct two sequences $(F^k)_k, (G^k)_k \subset L^2(B_1; \mathbb{R}^n)$, where $B_1 \subset \mathbb{R}^n$ is the unit ball, such that as $k \to \infty$

$$\begin{aligned} &F^k \to 0 && \text{weakly in } L^2(B_1; \mathbb{R}^n), && G^k \to 0 && \text{weakly in } L^2(B_1; \mathbb{R}^n),\\ &\operatorname{div} F^k = 0 && \text{in } W^{-1,2}(B_1), && \operatorname{curl} G^k = 0 && \text{in } W^{-1,2}(B_1; \mathbb{R}^{n\times n}), \end{aligned}$$

but as $k \to \infty$

$$\int_{B_{\frac{1}{2}}} F^k \cdot G^k \, \mathrm{d}x \not\to 0.$$

Thus, as $k \to \infty$

$$F^k \cdot G^k \not\to 0 = F^\infty \cdot G^\infty \quad \text{weakly in } L^1_{\mathrm{loc}}(B_1).$$

(d) A generalization of Theorem 1, where p' is replaced by an exponent $q \in (1, \infty)$ satisfying $1 \le 1/p + 1/q \le 1 + 1/n$, can be found in [4, Theorem 2.3]. Another generalization of Theorem 1, which assumes weaker conditions on $(\operatorname{div} F^k)_k$ and $(\operatorname{curl} G^k)_k$, but requires the equi-integrability of the sequence $(F^k \cdot G^k)_k$, is presented in [8]. Finally, for readers interested in the setting of Carnot groups, an analogue of Theorem 1 is proved in [2, Theorem 5.1].

To gain intuition about why this result holds, let us consider a simplified case where the vector fields G^k and G^∞ admit potentials, i.e., there exist scalar functions w^k and w^∞ such that $G^k = \nabla w^k$ and $G^\infty = \nabla w^\infty$. Then, $\operatorname{curl} G^k = \operatorname{curl} G^\infty = 0$, so that (3) trivially holds. In this case, the Div-Curl lemma can be proved more directly as a consequence of the integration by parts formula together with Rellich's compactness theorem.

Theorem 2 (Simplified Div-Curl Lemma) *Let* $(F^k)_k \subset L^p_{\mathrm{loc}}(\Omega;\mathbb{R}^n)$*,* $F^\infty \in L^p_{\mathrm{loc}}(\Omega;\mathbb{R}^n)$*,* $(w^k)_k \subset W^{1,p'}_{\mathrm{loc}}(\Omega)$*, and* $w^\infty \in W^{1,p'}_{\mathrm{loc}}(\Omega)$ *satisfy as* $k \to \infty$

$$F^k \to F^\infty \quad \text{weakly in } L^p_{\mathrm{loc}}(\Omega;\mathbb{R}^n), \qquad w^k \to w^\infty \quad \text{weakly in } W^{1,p'}_{\mathrm{loc}}(\Omega). \tag{5}$$

Assume that as $k \to \infty$

$$\operatorname{div} F^k \to \operatorname{div} F^\infty \quad \text{strongly in } W^{-1,p}_{\mathrm{loc}}(\Omega). \tag{6}$$

Then, as $k \to \infty$

$$F^k \cdot \nabla w^k \to F^\infty \cdot \nabla w^\infty \quad \text{weakly* in } \mathcal{D}'(\Omega). \tag{7}$$

Proof We fix a function $\varphi \in C^\infty_c(\Omega)$ and a bounded open set $\Omega' \subset \mathbb{R}^n$ with smooth boundary such that $\operatorname{supp}\varphi \subset \Omega' \Subset \Omega$.

For all $k \in \mathbb{N}$ we have

$$\begin{aligned}\int_\Omega \varphi F^k \cdot \nabla w^k \,\mathrm{d}x &= \int_{\Omega'} F^k \cdot \nabla(\varphi w^k)\,\mathrm{d}x - \int_{\Omega'} F^k \cdot \nabla\varphi w^k \,\mathrm{d}x\\ &= -\langle \operatorname{div} F^k, \varphi w^k\rangle_{W^{-1,p}(\Omega')\times W^{1,p'}_0(\Omega')} - \int_{\Omega'} F^k \cdot \nabla\varphi w^k \,\mathrm{d}x.\end{aligned}$$

Thanks to (5) and Rellich's theorem, as $k \to \infty$ we derive

$$\varphi w^k \to \varphi w^\infty \quad \text{weakly in } W^{1,p'}_0(\Omega'), \qquad w^k \to w^\infty \quad \text{strongly in } L^{p'}(\Omega'). \tag{8}$$

By combining (5)–(6) with (8), we get

$$\begin{aligned}\lim_{k\to\infty} \langle \operatorname{div} F^k, \varphi w^k\rangle_{W^{-1,p}(\Omega')\times W^{1,p'}_0(\Omega')} &= \langle \operatorname{div} F^\infty, \varphi w^\infty\rangle_{W^{-1,p}(\Omega')\times W^{1,p'}_0(\Omega')},\\ \lim_{k\to\infty}\int_{\Omega'} F^k \cdot \nabla\varphi w^k \,\mathrm{d}x &= \int_{\Omega'} F^\infty \cdot \nabla\varphi w^\infty \,\mathrm{d}x.\end{aligned}$$

Therefore,

$$\begin{aligned}\lim_{k\to\infty}\int_\Omega \varphi F^k \cdot \nabla w^k \,\mathrm{d}x = &-\langle \operatorname{div} F^\infty, \varphi w^\infty\rangle_{W^{-1,p}(\Omega')\times W^{1,p'}_0(\Omega')}\\ &+ \int_{\Omega'} F^\infty \cdot \nabla\varphi w^\infty \,\mathrm{d}x\end{aligned}$$

$$= \int_{\Omega'} F^\infty \cdot \nabla(\varphi w^\infty)\,\mathrm{d}x - \int_{\Omega'} F^\infty \cdot \nabla\varphi w^\infty\,\mathrm{d}x$$
$$= \int_{\Omega} \varphi F^\infty \cdot \nabla w^\infty\,\mathrm{d}x,$$

which gives (7). □

Remark 3 The simplified formulation of the Div-Curl lemma in Theorem 2 is sufficient for some applications, for example, in proving the compactness of nonlinear differential operators with oscillating coefficients, or the weak continuity of the determinant of the Jacobian matrix. One advantage of this version is that it avoids the explicit use of the curl operator, making it adaptable to non-Euclidean settings more general than Carnot groups (see [15, Theorem 3.1]).

Let us return to the general form of the Div-Curl lemma in Theorem 1. The key idea is to apply a Helmholtz decomposition to the sequence $(G^k)_k$ and its limit G^∞, writing

$$G^k = H^k + \nabla w^k \quad \text{for all } k \in \mathbb{N}, \qquad G^\infty = H^\infty + \nabla w^\infty,$$

where, as $k \to \infty$,

$$H^k \to H^\infty \quad \text{strongly in } L^{p'}_{\mathrm{loc}}(\Omega; \mathbb{R}^n), \qquad w^k \to w^\infty \quad \text{weakly in } W^{1,p'}_{\mathrm{loc}}(\Omega).$$

To construct such a decomposition, we recall some local regularity results for the distributional solutions $u \in L^1_{\mathrm{loc}}(\Omega)$ to the Poisson equation

$$-\Delta u = f \quad \text{in } \mathcal{D}'(\Omega),$$

where the right-hand side f belongs either to $W^{-1,p}_{\mathrm{loc}}(\Omega)$ or to $L^p_{\mathrm{loc}}(\Omega)$. These results are quite classical (see, e.g., [11]), but for the reader's convenience we provide a simple proof following a new strategy introduced in [9]. The key ingredient is the following proposition, which can be interpreted as a $W^{1,p}$-version of the classical Riesz representation theorem.

Proposition 1 ([21, Theorem 1.2, Chapter 2]) *Let $\Omega \subset \mathbb{R}^n$ be a bounded open set with boundary $\partial\Omega$ of class C^1. For every $f \in W^{-1,p}(\Omega)$, there exists a unique function $v \in W^{1,p}_0(\Omega)$ satisfying*

$$\int_\Omega \nabla v \cdot \nabla \phi\,\mathrm{d}x = \langle f, \phi \rangle_{W^{-1,p}(\Omega) \times W^{1,p'}_0(\Omega)} \quad \textit{for all } \phi \in W^{1,p'}_0(\Omega).$$

Moreover, there exists a constant $C = C(n, p, \Omega) > 0$ such that

$$\|v\|_{W^{1,p}(\Omega)} \le C \|f\|_{W^{-1,p}(\Omega)}. \tag{9}$$

Remark 4 Actually, in [21, Theorem 1.2, Chapter 2], the authors also assume that Ω is connected. However, this assumption can be omitted by observing that any bounded open set $\Omega \subset \mathbb{R}^n$ with boundary $\partial\Omega$ of class C^1 has only a finite number of connected components. Therefore, it is sufficient to apply [21, Theorem 1.2, Chapter 2] to each connected component to obtain Proposition 1. Furthermore, the estimate in (9) follows from the one in [21, Theorem 1.2, Chapter 2] by applying Poincaré's inequality.

As a consequence of Proposition 1, we can argue as in [9, Theorem 3] to derive the following two local regularity results.

Proposition 2 *Let $f \in W^{-1,p}_{\mathrm{loc}}(\Omega)$ and $u \in L^1_{\mathrm{loc}}(\Omega)$ be a solution to*

$$-\Delta u = f \quad \text{in } \mathcal{D}'(\Omega).$$

Then $u \in W^{1,p}_{\mathrm{loc}}(\Omega)$ and for all bounded open sets $\Omega'' \Subset \Omega' \Subset \Omega$ there exists a constant $C = C(n, p, \Omega', \Omega'') > 0$ such that

$$\|u\|_{W^{1,p}(\Omega'')} \le C(\|u\|_{L^1(\Omega')} + \|f\|_{W^{-1,p}(\Omega')}). \tag{10}$$

Proof We fix two bounded open sets $\Omega'' \Subset \Omega' \Subset \Omega$. Without loss of generality, we may assume that Ω' has a smooth boundary (otherwise, we replace Ω' with a bounded open set Ω''' with smooth boundary such that $\Omega'' \Subset \Omega''' \Subset \Omega'$). To complete the proof, it is enough to show that $u \in W^{1,p}(\Omega'')$ and that the estimate (10) holds.

Since $f \in W^{-1,p}(\Omega')$, by Proposition 1 there is a function $v \in W^{1,p}_0(\Omega')$ satisfying

$$-\Delta v = f \quad \text{in } \mathcal{D}'(\Omega'),$$

and we can find a constant $C = C(n, p, \Omega') > 0$ such that

$$\|v\|_{W^{1,p}(\Omega')} \le C\|f\|_{W^{-1,p}(\Omega')}. \tag{11}$$

The difference $u - v \in L^1(\Omega')$ is a harmonic distribution in Ω', so by Weyl's lemma (see, e.g., [13, Lemma 1.16]), there exists a harmonic function $z \in C^\infty(\Omega')$ such that $z = u - v$ in Ω'. Therefore, $u = z + v \in W^{1,p}(\Omega'')$, and from (11) it follows that

$$\|u\|_{W^{1,p}(\Omega'')} \le \|z\|_{W^{1,p}(\Omega'')} + \|v\|_{W^{1,p}(\Omega'')} \le \|z\|_{W^{1,p}(\Omega'')} + \|f\|_{W^{-1,p}(\Omega')}. \tag{12}$$

Since z is harmonic in Ω', Harnack's and Caccioppoli's inequalities (see, e.g., [11, Theorem 7, Section 2.2]) imply the existence of a constant $C = C(n, p, \Omega', \Omega'') > 0$ such that

$$\|z\|_{W^{1,p}(\Omega'')} \leq C\|z\|_{L^1(\Omega')} \leq C(\|u\|_{L^1(\Omega')} + \|v\|_{L^1(\Omega')}). \tag{13}$$

By combining (11)–(13), we obtain (10). □

Proposition 3 *Let $f \in L^p_{\text{loc}}(\Omega)$ and $u \in L^1_{\text{loc}}(\Omega)$ be a solution to*

$$-\Delta u = f \quad \textit{in } \mathcal{D}'(\Omega).$$

Then $u \in W^{2,p}_{\text{loc}}(\Omega)$ and for all bounded open sets $\Omega'' \Subset \Omega' \Subset \Omega$ there exists a constant $C = C(n, p, \Omega', \Omega'') > 0$ such that

$$\|u\|_{W^{2,p}(\Omega'')} \leq C(\|u\|_{L^1(\Omega')} + \|f\|_{L^p(\Omega')}). \tag{14}$$

Proof We fix two bounded open sets $\Omega'' \Subset \Omega' \Subset \Omega$.

Since $f \in L^p_{\text{loc}}(\Omega) \subset W^{-1,p}_{\text{loc}}(\Omega)$, by Proposition 2 we derive that $u \in W^{1,p}_{\text{loc}}(\Omega)$ and that there exists a constant $C = C(n, p, \Omega', \Omega'') > 0$ such that

$$\|u\|_{W^{1,p}(\Omega'')} \leq C(\|u\|_{L^1(\Omega')} + \|f\|_{W^{-1,p}(\Omega')}) \leq C(\|u\|_{L^1(\Omega')} + \|f\|_{L^p(\Omega')}). \tag{15}$$

Moreover, $\nabla u \in L^p_{\text{loc}}(\Omega; \mathbb{R}^n)$ is a distributional solution to

$$-\Delta(\nabla u) = \nabla f \quad \text{in } \mathcal{D}'(\Omega; \mathbb{R}^n),$$

with $\nabla f \in W^{-1,p}_{\text{loc}}(\Omega; \mathbb{R}^n)$. By applying again Proposition 2 we deduce that $\nabla u \in W^{1,p}_{\text{loc}}(\Omega; \mathbb{R}^n)$ and that there exists a constant $C = C(n, p, \Omega', \Omega'') > 0$ such that

$$\begin{aligned}\|\nabla u\|_{W^{1,p}(\Omega'';\mathbb{R}^n)} &\leq C(\|\nabla u\|_{L^1(\Omega';\mathbb{R}^n)} + \|\nabla f\|_{W^{-1,p}(\Omega';\mathbb{R}^n)}) \\ &\leq C(\|\nabla u\|_{L^1(\Omega';\mathbb{R}^n)} + \|f\|_{L^p(\Omega')}).\end{aligned} \tag{16}$$

Hence, $u \in W^{2,p}_{\text{loc}}(\Omega)$ and by combining (15) and (16) we derive (14). □

We are now in a position to prove Theorem 1, following the strategy adopted in [10, Theorem 4, Chapter 5] for the case $p = 2$, see also [16, Théorème 2].

Proof of Theorem 1 We fix a function $\varphi \in C^\infty_c(\Omega)$ and a bounded open set Ω' with smooth boundary such that $\operatorname{supp}\varphi \subset \Omega' \Subset \Omega$.

For all $k \in \mathbb{N}$ we consider the function $V^k \in W^{1,p'}_0(\Omega'; \mathbb{R}^n)$ given by Proposition 1, which solves

$$-\Delta V^k = G^k - G^\infty \quad \text{in } \mathcal{D}'(\Omega'; \mathbb{R}^n). \tag{17}$$

In particular, by using also Remark 1, there exists a constant $C = C(n, p, \Omega) > 0$, independent of $k \in \mathbb{N}$, such that for all $k \in \mathbb{N}$

$$\|V^k\|_{W^{1,p'}(\Omega';\mathbb{R}^n)} \leq C\|G^k - G^\infty\|_{W^{-1,p'}(\Omega';\mathbb{R}^n)} \leq C\|G^k - G^\infty\|_{L^{p'}(\Omega';\mathbb{R}^n)}.$$

Therefore, the sequence $(V^k)_k \subset W_0^{1,p'}(\Omega';\mathbb{R}^n)$ is uniformly bounded, and by (1) and (17) we derive that as $k \to \infty$

$$V^k \to 0 \quad \text{weakly in } W_0^{1,p'}(\Omega';\mathbb{R}^n). \tag{18}$$

Thanks to the local regularity result of Proposition 3, we derive that $V^k \in W^{2,p'}_{\mathrm{loc}}(\Omega';\mathbb{R}^n)$ and for all bounded open sets $\Omega'' \Subset \Omega'$ we can find a constant $C = C(n, p, \Omega', \Omega'') > 0$, independent of $k \in \mathbb{N}$, such that for all $k \in \mathbb{N}$

$$\|V^k\|_{W^{2,p'}(\Omega'';\mathbb{R}^n)} \leq C(\|V^k\|_{L^1(\Omega';\mathbb{R}^n)} + \|G^k - G^\infty\|_{L^{p'}(\Omega';\mathbb{R}^n)}).$$

In particular, the sequence $(V^k)_k \subset W^{2,p'}_{\mathrm{loc}}(\Omega';\mathbb{R}^n)$ is uniformly bounded. By (18) and Rellich's theorem, we conclude that as $k \to \infty$

$$V^k \to 0 \quad \text{weakly in } W^{2,p'}_{\mathrm{loc}}(\Omega';\mathbb{R}^n), \qquad V^k \to 0 \quad \text{strongly in } W^{1,p'}_{\mathrm{loc}}(\Omega';\mathbb{R}^n). \tag{19}$$

The function $\operatorname{curl} V^k \in W^{1,p'}_{\mathrm{loc}}(\Omega';\mathbb{R}^{n\times n})$ satisfies for all $k \in \mathbb{N}$

$$-\Delta \operatorname{curl} V^k = \operatorname{curl} G^k - \operatorname{curl} G^\infty \quad \text{in } \mathcal{D}'(\Omega';\mathbb{R}^{n\times n}).$$

Therefore, by Proposition 2 for all bounded open sets $\Omega''' \Subset \Omega'' \Subset \Omega'$ there is $C = C(n, p, \Omega'', \Omega''') > 0$, independent of $k \in \mathbb{N}$, such that for all $k \in \mathbb{N}$

$$\begin{aligned}&\|\operatorname{curl} V^k\|_{W^{1,p'}(\Omega''';\mathbb{R}^{n\times n})}\\ &\leq C(\|\operatorname{curl} V^k\|_{L^1(\Omega'';\mathbb{R}^{n\times n})} + \|\operatorname{curl} G^k - \operatorname{curl} G^\infty\|_{W^{-1,p'}(\Omega'';\mathbb{R}^{n\times n})}).\end{aligned}$$

Hence, by (3) and (19), as $k \to \infty$ we have

$$\operatorname{curl} V^k \to 0 \quad \text{strongly in } W^{1,p'}_{\mathrm{loc}}(\Omega';\mathbb{R}^{n\times n}). \tag{20}$$

Let us define

$$w^k := -\operatorname{div} V^k \quad \text{for all } k \in \mathbb{N}, \qquad H^k := G^k - G^\infty - \nabla w^k \quad \text{for all } k \in \mathbb{N}. \tag{21}$$

By (19) we have that $(w^k)_k \subset W^{1,p'}_{\mathrm{loc}}(\Omega')$ and as $k \to \infty$

$$w^k \to 0 \quad \text{weakly in } W^{1,p'}_{\mathrm{loc}}(\Omega'), \qquad w^k \to 0 \quad \text{strongly in } L^{p'}_{\mathrm{loc}}(\Omega'). \tag{22}$$

Moreover, $(H^k)_k \subset L^{p'}_{\mathrm{loc}}(\Omega'; \mathbb{R}^n)$ and

$$H^k = -\operatorname{div}(\operatorname{curl} V^k) \quad \text{in } \mathcal{D}'(\Omega'; \mathbb{R}^n) \text{ for all } k \in \mathbb{N}.$$

Indeed, for all $k \in \mathbb{N}$ and $i \in \{1, \dots, n\}$ we have

$$\begin{aligned} H_i^k &= G_i^k - G_i^\infty - \partial_i w^k = -\Delta V_i^k + \partial_i \operatorname{div} V^k \\ &= -\sum_{j=1}^n \partial_j(\partial_j V_i^k - \partial_i V_j^k) = -\sum_{j=1}^n \partial_j (\operatorname{curl} V^k)_{ij} = -(\operatorname{div}(\operatorname{curl} V^k))_i \text{ in } \mathcal{D}'(\Omega'). \end{aligned}$$

Therefore, by (20) we derive that as $k \to \infty$

$$H^k \to 0 \quad \text{strongly in } L^{p'}_{\mathrm{loc}}(\Omega'; \mathbb{R}^n). \tag{23}$$

In view of (21), we can write

$$\int_\Omega \varphi F^k \cdot G^k \, \mathrm{d}x = \int_{\Omega'} \varphi F^k \cdot G^\infty \, \mathrm{d}x + \int_{\Omega'} \varphi F^k \cdot H^k \, \mathrm{d}x + \int_{\Omega'} \varphi F^k \cdot \nabla w^k \, \mathrm{d}x.$$

Thanks to (1), (2), and (22), we can apply Theorem 2 to deduce that

$$\lim_{k\to\infty} \int_{\Omega'} \varphi F^k \cdot \nabla w^k \, \mathrm{d}x = 0.$$

Moreover, by (1) and (23) we get

$$\lim_{k\to\infty} \int_{\Omega'} \varphi F^k \cdot G^\infty \, \mathrm{d}x = \int_{\Omega'} \varphi F^\infty \cdot G^\infty \, \mathrm{d}x, \qquad \lim_{k\to\infty} \int_{\Omega'} \varphi F^k \cdot H^k \, \mathrm{d}x = 0.$$

Hence, for all $\varphi \in C_c^\infty(\Omega)$ we deduce

$$\lim_{k\to\infty} \int_\Omega \varphi F^k \cdot G^k \, \mathrm{d}x = \int_\Omega \varphi F^\infty \cdot G^\infty \, \mathrm{d}x,$$

which gives (4). □

3 Some Extensions to Fractional Sobolev Spaces

In this section, we present several generalizations of the Div-Curl lemma to the fractional setting of [6, 18]. We begin by briefly recalling the definitions of the Riesz fractional gradient and divergence, along with the key properties of the associated fractional Sobolev spaces. We then state and prove three generalizations of the Div-Curl lemma. The first, Theorem 3, extends Theorem 2 to this fractional framework. The second, Theorem 4, is a mixed local-nonlocal version, where we assume convergence of the fractional divergence for the first sequence and standard curl convergence for the second. The third, Theorem 5, is another mixed local-nonlocal result, in the simplified setting of Theorem 2, where we replace the gradient with its fractional counterpart, while we keep the convergence of the standard divergence.

3.1 The Fractional Framework

In this subsection, we introduce the fractional operators that will be considered throughout this chapter and collect all the results that will be used in the proofs of the fractional Div-Curl lemmas.

Let $n \geq 2$, let $p \in (1, \infty)$, and let $s \in (0, 1)$ be fixed. We define

$$\mu_{n,s} := \frac{2^s \Gamma(\frac{n+s+1}{2})}{\pi^{\frac{n}{2}} \Gamma(\frac{1-s}{2})} \in (0, \infty),$$

where Γ denotes the varGamma function.

Definition 2 Given a function $\psi \in C_c^\infty(\mathbb{R}^n)$, we define the *Riesz s-fractional gradient* $\nabla^s \psi : \mathbb{R}^n \to \mathbb{R}^n$ by

$$\nabla^s \psi(x) := \mu_{n,s} \int_{\mathbb{R}^n} \frac{(\psi(y) - \psi(x))(y - x)}{|y - x|^{n+s+1}} \, dy \quad \text{for all } x \in \mathbb{R}^n.$$

Given a function $\Psi \in C_c^\infty(\mathbb{R}^n; \mathbb{R}^n)$, we define the *Riesz s-fractional divergence* $\operatorname{div}^s \Psi : \mathbb{R}^n \to \mathbb{R}$ by

$$\operatorname{div}^s \Psi(x) := \mu_{n,s} \int_{\mathbb{R}^n} \frac{(\Psi(y) - \Psi(x)) \cdot (y - x)}{|y - x|^{n+s+1}} \, dy \quad \text{for all } x \in \mathbb{R}^n.$$

As observed in [6, Section 2], the nonlocal operators $\nabla^s \psi$ and $\operatorname{div}^s \Psi$ are well-defined in the sense that the above integrals converge for all $x \in \mathbb{R}^n$. Moreover, these operators satisfy the following L^p-type estimates, which can be found in [7, Lemmas 2.2-2.3 and Propositions 3.2-3.3].

Proposition 4 *There exists a constant $C = C(n, s) > 0$ such that for all $q \in [1, \infty]$ and for all $\psi \in C_c^\infty(\mathbb{R}^n)$ and $\Psi \in C_c^\infty(\mathbb{R}^n; \mathbb{R}^n)$ we have*

$$\|\nabla^s \psi\|_{L^q(\mathbb{R}^n;\mathbb{R}^n)} \le C\|\psi\|_{L^q(\mathbb{R}^n)}^{1-s}\|\nabla\psi\|_{L^q(\mathbb{R}^n;\mathbb{R}^n)}^s,$$
$$\|\operatorname{div}^s \Psi\|_{L^q(\mathbb{R}^n)} \le C\|\Psi\|_{L^q(\mathbb{R}^n;\mathbb{R}^n)}^{1-s}\|\nabla\Psi\|_{L^q(\mathbb{R}^n;\mathbb{R}^{n\times n})}^s.$$

As shown in [6, Lemma 2.5] and [20, Section 6], the nonlocal operators ∇^s and div^s satisfy the following integration by parts formula.

Proposition 5 *For all $\psi \in C_c^\infty(\mathbb{R}^n)$ and $\Psi \in C_c^\infty(\mathbb{R}^n; \mathbb{R}^n)$ we have*

$$\int_{\mathbb{R}^n} \nabla^s \psi \cdot \Psi \,\mathrm{d}x = -\int_{\mathbb{R}^n} \psi \operatorname{div}^s \Psi \,\mathrm{d}x. \tag{24}$$

Remark 5 These two fractional operators are closely connected to the classical *fractional Laplacian* $(-\Delta)^s$, which for $\psi \in C_c^\infty(\mathbb{R}^n)$ and $x \in \mathbb{R}^n$ is defined by

$$(-\Delta)^s \psi(x) := \begin{cases} \nu_{n,s} \displaystyle\int_{\mathbb{R}^n} \frac{\psi(y) - \psi(x)}{|x-y|^{n+2s}} \,\mathrm{d}y & \text{if } s \in (0, \tfrac{1}{2}), \\ \nu_{n,s} \displaystyle\lim_{\varepsilon \to 0^+} \int_{\{y \in \mathbb{R}^n \,:\, |x-y| > \varepsilon\}} \frac{\psi(y) - \psi(x)}{|x-y|^{n+2s}} \,\mathrm{d}y & \text{if } s \in [\tfrac{1}{2}, 1), \end{cases}$$

where

$$\nu_{n,s} := \frac{2^{2s}\Gamma(\frac{n+2s}{2})}{\pi^{\frac{n}{2}}\Gamma(-s)} \in (-\infty, 0).$$

Indeed, as observed in [20, Theorem 5.3], the following relation holds for all functions $\psi \in C_c^\infty(\mathbb{R}^n)$ and $s, r \in (0, 1)$

$$-\operatorname{div}^s \nabla^r \psi = (-\Delta)^{\frac{s+r}{2}} \psi \quad \text{in } \mathbb{R}^n. \tag{25}$$

Let $\Omega \subset \mathbb{R}^n$ be an open set. We now introduce the fractional Sobolev framework for our Div-Curl lemmas.

Definition 3 Given a function $\psi \in C_c^\infty(\mathbb{R}^n)$, we define the norm

$$\|\psi\|_{H^{s,p}(\mathbb{R}^n)} := (\|\psi\|_{L^p(\mathbb{R}^n)}^p + \|\nabla^s \psi\|_{L^p(\mathbb{R}^n;\mathbb{R}^n)}^p)^{1/p},$$

and the fractional Sobolev spaces $H_0^{s,p}(\Omega)$ and $H^{-s,p}(\Omega)$, respectively, by

$$H_0^{s,p}(\Omega) := \overline{C_c^\infty(\Omega)}^{\|\cdot\|_{H^{s,p}(\mathbb{R}^n)}}, \qquad H^{-s,p}(\Omega) := (H_0^{s,p'}(\Omega))'.$$

When $\Omega = \mathbb{R}^n$, we simply write $H^{s,p}(\mathbb{R}^n) := H_0^{s,p}(\mathbb{R}^n)$. Finally, we define the local negative fractional Sobolev space

$$H_{\mathrm{loc}}^{-s,p}(\Omega) := \{f \in \mathcal{D}'(\Omega) \,:\, f \in H^{-s,p}(\Omega') \text{ for all open sets } \Omega' \Subset \Omega\}.$$

Remark 6 We denote the above fractional Sobolev space by $H^{s,p}$ instead of $W^{s,p}$, as the latter is typically used for the fractional Sobolev-Slobodeckij space defined via Gagliardo seminorms. These two spaces generally differ, except in the case $p = 2$, see [18, Theorems 1.7 and 2.2]. In particular, when $p = 2$, there exists a constant $C = C(n, s) > 0$ such that for all $u \in H^{s,2}(\mathbb{R}^n)$ we have

$$\|\nabla^s u\|_{L^2(\mathbb{R}^n;\mathbb{R}^n)}^2 = C \int_{\mathbb{R}^n} \int_{\mathbb{R}^n} \frac{|u(y) - u(x)|^2}{|y - x|^{n+2s}} \, dx \, dy,$$

see, for instance, [1, Proposition 2.8].

If we extend the operators ∇^s and div^s to $H^{s,p}(\mathbb{R}^n)$ and $H^{s,p'}(\mathbb{R}^n;\mathbb{R}^n)$, respectively, then the integration by parts formula (24) holds for all functions $\psi \in H^{s,p}(\mathbb{R}^n)$ and $\Psi \in H^{s,p'}(\mathbb{R}^n;\mathbb{R}^n)$. This justify the following definition.

Definition 4 Given a vector field $F \in L^p(\mathbb{R}^n;\mathbb{R}^n)$, we define $\mathrm{div}^s F \in H_{\mathrm{loc}}^{-s,p}(\Omega)$ for all bounded open sets $\Omega' \Subset \Omega$ by

$$\langle \mathrm{div}^s F, v\rangle_{H^{-s,p}(\Omega')\times H_0^{s,p'}(\Omega)} := -\int_{\mathbb{R}^n} F \cdot \nabla^s v \, dx \quad \text{for all } v \in H_0^{s,p'}(\Omega').$$

We recall the following result, which provide a useful connection between the Sobolev spaces $W^{1,p}(\mathbb{R}^n)$ and $H^{s,p}(\mathbb{R}^n)$. For the proof, we refer the interested reader to [14, Proposition 3.1].

Proposition 6 *The following two statements hold.*

(a) For all functions $v \in H^{s,p}(\mathbb{R}^n)$ there exists a function $u \in W_{\mathrm{loc}}^{1,p}(\mathbb{R}^n)$ such that

$$\nabla u = \nabla^s v \quad \text{in } \mathbb{R}^n.$$

(b) For all functions $u \in W^{1,p}(\mathbb{R}^n)$ there exists a function $v \in H^{s,p}(\mathbb{R}^n)$ such that

$$\nabla^s v = \nabla u \quad \text{in } \mathbb{R}^n.$$

Moreover, there exists a constant $C = C(n, s) > 0$ such that

$$\|v\|_{L^p(\mathbb{R}^n)} \leq C\|u\|_{L^p(\mathbb{R}^n)}^{1-s} \|\nabla u\|_{L^p(\mathbb{R}^n;\mathbb{R}^n)}^s. \tag{26}$$

Remark 7 If we recall the fractional Laplacian defined in Remark 5, then in Proposition 6(b) we have that $v = (-\Delta)^{\frac{1-s}{2}} u$. On the other hand, the function v in Proposition 6(a) is constructed by using the inverse of the fractional Laplacian, which is the *Riesz potential.*

We conclude this subsection by recalling a Leibniz-type rule for the fractional gradient. To this aim, we need to introduce the following nonlocal operator.

Definition 5 For all $\psi, \phi \in C_c^\infty(\mathbb{R}^n)$ we define $\nabla_{\mathrm{NL}}^s(\psi, \phi) : \mathbb{R}^n \to \mathbb{R}^n$ for all $x \in \mathbb{R}^n$ by

$$\nabla_{\mathrm{NL}}^s(\psi, \phi)(x) := \mu_{n,s} \int_{\mathbb{R}^n} \frac{(\psi(y) - \psi(x))(\phi(y) - \phi(x))(y - x)}{|y - x|^{n+s+1}} \, dy \tag{27}$$

As before, the integral (27) converges for all $x \in \mathbb{R}^n$, and the following L^p-type estimate hold, see [14, Eq. (2.11)].

Proposition 7 *There exists a constant $C = C(n, s) > 0$ such that for all functions $\psi, \phi \in C_c^\infty(\mathbb{R}^n)$ we have*

$$\|\nabla_{\mathrm{NL}}^s(\psi, \phi)\|_{L^p(\mathbb{R}^n;\mathbb{R}^n)} \le C\|\psi\|_{L^p(\mathbb{R}^n)}\|\phi\|_{L^\infty(\mathbb{R}^n)}^s \|\nabla\phi\|_{L^\infty(\mathbb{R}^n;\mathbb{R}^n)}^{1-s}. \tag{28}$$

In view of (28), the nonlocal operator $\nabla_{\mathrm{NL}}(\psi, \phi)$ can be extended to all pairs $(\psi, \phi) \in L^p(\mathbb{R}^n) \times C_c^\infty(\mathbb{R}^n)$. In particular, we have the following Leibniz-type rule for the fractional gradient, see [14, Lemma 2.11] for a proof.

Proposition 8 *Let $\phi \in C_c^\infty(\Omega)$ and $u \in H^{s,p}(\mathbb{R}^n)$. Then, $u\phi \in H_0^{s,p}(\Omega)$ and*

$$\nabla^s(u\phi)(x) = u(x)\nabla^s\phi(x) + \phi(x)\nabla^s u(x) + \nabla_{\mathrm{NL}}^s(u, \phi)(x). \tag{29}$$

Moreover, there exists a constant $C = C(n, s) > 0$ such that

$$\|\nabla_{\mathrm{NL}}^s(u, \phi)\|_{L^p(\mathbb{R}^n;\mathbb{R}^n)} \le C\|u\|_{L^p(\mathbb{R}^n)}\|\phi\|_{L^\infty(\mathbb{R}^n)}^s \|\nabla\phi\|_{L^\infty(\mathbb{R}^n;\mathbb{R}^n)}^{1-s}. \tag{30}$$

For further properties of the Riesz fractional gradient and divergence, as well as of the fractional Sobolev space $H^{s,p}$, we refer the reader to [6, 7, 18–20].

3.2 Three Fractional Div-Curl Lemmas

We can finally present three possible extensions of the Div-Curl lemma to the fractional setting introduced earlier. We begin with a fractional version of Theorem 2.

Theorem 3 (Simplified Fractional Div-Curl Lemma) *Let* $(F^k)_k \subset L^p(\mathbb{R}^n;\mathbb{R}^n)$, $F^\infty \in L^p(\mathbb{R}^n;\mathbb{R}^n)$, $(w^k)_k \subset H^{s,p'}(\mathbb{R}^n)$, *and* $w^\infty \in H^{s,p'}(\mathbb{R}^n)$ *satisfy as* $k \to \infty$

$$F^k \to F^\infty \qquad \text{weakly in } L^p(\mathbb{R}^n;\mathbb{R}^n), \tag{31}$$

$$\nabla^s w^k \to \nabla^s w^\infty \qquad \text{weakly in } L^{p'}_{\mathrm{loc}}(\Omega;\mathbb{R}^n). \tag{32}$$

Assume that

$$\operatorname{div}^s F^k \to \operatorname{div}^s F^\infty \qquad \text{strongly in } H^{-s,p}_{\mathrm{loc}}(\Omega), \tag{33}$$

$$w^k \to w^\infty \qquad \text{strongly in } L^{p'}(\mathbb{R}^n). \tag{34}$$

Then, as $k \to \infty$

$$F^k \cdot \nabla^s w^k \to F^\infty \cdot \nabla^s w^\infty \quad \text{weakly* in } \mathcal{D}'(\Omega). \tag{35}$$

Proof We fix a function $\varphi \in C_c^\infty(\Omega)$ and a bounded open set $\Omega' \subset \mathbb{R}^n$ with smooth boundary such that $\operatorname{supp}\varphi \subset \Omega' \Subset \Omega$.

By Proposition 8 we have

$$\int_{\mathbb{R}^n} \varphi F^k \cdot \nabla^s w^k \,\mathrm{d}x = \int_{\mathbb{R}^n} F^k \cdot \nabla^s(\varphi w^k)\,\mathrm{d}x - \int_{\mathbb{R}^n} F^k \cdot \nabla^s \varphi w^k \,\mathrm{d}x$$
$$- \int_{\mathbb{R}^n} F^k \cdot \nabla^s_{\mathrm{NL}}(w^k,\varphi)\,\mathrm{d}x. \tag{36}$$

Since $\nabla^s\varphi \in L^\infty(\mathbb{R}^n;\mathbb{R}^n)$ by Proposition 4, the weak convergence (31) and the strong convergence in (34) imply

$$\lim_{k\to\infty} \int_{\mathbb{R}^n} F^k \cdot \nabla^s \varphi w^k \,\mathrm{d}x = \int_{\mathbb{R}^n} F^\infty \cdot \nabla^s \varphi w^\infty \,\mathrm{d}x. \tag{37}$$

By the estimate (30), the strong convergence in (34), and the fact that $\nabla^s_{\mathrm{NL}}(\cdot\,,\cdot)$ is a bilinear operator, we deduce that as $k \to \infty$

$$\nabla^s_{\mathrm{NL}}(w^k,\varphi) \to \nabla^s_{\mathrm{NL}}(w^\infty,\varphi) \quad \text{strongly in } L^{p'}(\mathbb{R}^n;\mathbb{R}^n).$$

Thus, in view of (31), we conclude that

$$\lim_{k\to\infty} \int_{\mathbb{R}^n} F^k \cdot \nabla^s_{\mathrm{NL}}(w^k,\varphi)\,\mathrm{d}x = \int_{\mathbb{R}^n} F^\infty \cdot \nabla^s_{\mathrm{NL}}(w^\infty,\varphi)\,\mathrm{d}x. \tag{38}$$

Finally, by Proposition 8 we have that $\varphi w^k \in H_0^{s,p'}(\Omega')$. Moreover, in view of (29) and (32), as $k \to \infty$ we derive

$$\varphi w^k \to \varphi w^\infty \quad \text{weakly in } H_0^{s,p'}(\Omega').$$

Hence, by using also (33), we have

$$\begin{aligned}\lim_{k\to\infty}\int_{\mathbb{R}^n} F^k\cdot\nabla^s(\varphi w^k)\,\mathrm{d}x &= -\lim_{k\to\infty}\langle\operatorname{div}^s F^k,\varphi w^k\rangle_{H^{-s,p}(\Omega')\times H_0^{s,p'}(\Omega')}\\ &= -\langle\operatorname{div}^s F^\infty,\varphi w^\infty\rangle_{H^{-s,p}(\Omega')\times H_0^{s,p'}(\Omega')}\\ &= \int_{\mathbb{R}^n} F^\infty\cdot\nabla^s(\varphi w^\infty)\,\mathrm{d}x. \end{aligned}\tag{39}$$

By combining (36)–(39), for all $\varphi\in C_c^\infty(\Omega)$ we obtain

$$\begin{aligned}\lim_{k\to\infty}\int_{\mathbb{R}^n}\varphi F^k\cdot\nabla^s w^k\,\mathrm{d}x &= \int_{\mathbb{R}^n} F^\infty\cdot\nabla^s(\varphi w^\infty)\,\mathrm{d}x - \int_{\mathbb{R}^n} F^\infty\cdot\nabla^s\varphi w^\infty\,\mathrm{d}x\\ &\quad-\int_{\mathbb{R}^n} F^\infty\cdot\nabla^s_{\mathrm{NL}}(w^\infty,\varphi)\,\mathrm{d}x\\ &= \int_{\mathbb{R}^n}\varphi F^\infty\cdot\nabla^s w^\infty\,\mathrm{d}x,\end{aligned}$$

which is (35). □

Remark 8 Let us highlight the main differences compared to the analogous result stated in Theorem 2.

(a) The first difference is that the sequences $(F^k)_k$ and $(w^k)_k$ must be defined on the whole space $\mathbb{R}^n$, rather than just on Ω. This requirement arises from the nonlocal nature of the fractional gradient: in order to define $\nabla^s\psi$ and $\operatorname{div}^s\Psi$ at a point $x\in\mathbb{R}^n$, one needs to know the values of ψ and Ψ on the entire space $\mathbb{R}^n$.
(b) Moreover, the weak convergence (31) must also hold on the whole space $\mathbb{R}^n$. This is a consequence of the integration by parts formula stated in Proposition 5, which involves integration over all of $\mathbb{R}^n$. In particular, the integrals in (39) must be over $\mathbb{R}^n$, which in turn forces all the integrals in (36) to be taken over $\mathbb{R}^n$ as well. Consequently, the sequence $(F^k)_k$ must converge to F^∞ weakly in $L^p(\mathbb{R}^n;\mathbb{R}^n)$.
(c) A further difference is that we must explicitly assume (34). In the local case, such strong convergence is only required on $\Omega'\Subset\Omega$, a bounded open set, and therefore follows from Rellich's theorem. In the nonlocal case, however, the strong convergence must hold globally on $\mathbb{R}^n$. We observe that, if $\Omega\subset\mathbb{R}^n$ is a bounded open set and $(w^k)_k\subset H_0^{s,p'}(\Omega)$, $w^\infty\in H_0^{s,p'}(\Omega)$ satisfy as $k\to\infty$

$$\nabla^s w^k\rightharpoonup\nabla^s w^\infty\quad\text{weakly in } L^{p'}(\mathbb{R}^n;\mathbb{R}^n),$$

then (34) follows directly from the fractional versions of the Poincaré inequality (see [18, Theorem 3.3]) and Rellich's compactness theorem (see [19, Theorem 2.2] and [3, Theorem 2.3]).

Remark 9 We point out to the interested reader that, in the context of fractional operators defined via Gagliardo-type double integrals, an analogue of Theorem 3 can be found in [12, Lemma 3.3]. In particular, this version is used to prove the compactness of a nonlinear Gagliardo-type differential operator depending on an oscillating scalar weight.

As a consequence of Theorem 3 and Proposition 6, we can derive the following "mixed local-nonlocal" Div-Curl lemma. In this result, the fractional divergence is assumed to converge in $H^{-s,p}_{\mathrm{loc}}(\Omega)$, while the classical curl is required to converge in $W^{-1,p'}_{\mathrm{loc}}(\Omega;\mathbb{R}^{n\times n})$.

Theorem 4 (First Mixed Local–Nonlocal Div-Curl Lemma) *Let* $(F^k)_k \subset L^p(\mathbb{R}^n;\mathbb{R}^n)$, $F^\infty \in L^p(\mathbb{R}^n;\mathbb{R}^n)$, $(G^k)_k \subset L^{p'}_{\mathrm{loc}}(\Omega;\mathbb{R}^n)$, *and* $G^\infty \in L^{p'}_{\mathrm{loc}}(\Omega;\mathbb{R}^n)$ *satisfy as* $k \to \infty$

$$F^k \to F^\infty \quad \textit{weakly in } L^p(\mathbb{R}^n;\mathbb{R}^n), \qquad G^k \to G^\infty \quad \textit{weakly in } L^{p'}_{\mathrm{loc}}(\Omega;\mathbb{R}^n). \tag{40}$$

Assume that as $k \to \infty$

$$\begin{aligned} \operatorname{div}^s F^k &\to \operatorname{div}^s F^\infty && \textit{strongly in } H^{-s,p}_{\mathrm{loc}}(\Omega), \\ \operatorname{curl} G^k &\to \operatorname{curl} G^\infty && \textit{strongly in } W^{-1,p'}_{\mathrm{loc}}(\Omega;\mathbb{R}^{n\times n}). \end{aligned}$$

Then, as $k \to \infty$

$$F^k \cdot G^k \to F^\infty \cdot G^\infty \quad \textit{weakly* in } \mathcal{D}'(\Omega). \tag{41}$$

Proof We fix a function $\varphi \in C^\infty_c(\Omega)$ and a bounded open set $\Omega' \subset \mathbb{R}^n$ with smooth boundary such that $\operatorname{supp}\varphi \subset \Omega' \Subset \Omega$. We also fix a function $\psi \in C^\infty_c(\Omega')$ such that $\psi = 1$ on $\operatorname{supp}\varphi$.

Let $(w^k)_k \subset W^{1,p'}_{\mathrm{loc}}(\Omega')$ and $(H^k)_k \subset L^{p'}_{\mathrm{loc}}(\Omega';\mathbb{R}^n)$ be the functions defined in (21). Since $\psi = 1$ on $\operatorname{supp}\varphi$, by (21) we have

$$\begin{aligned} \int_\Omega \varphi F^k \cdot G^k \,\mathrm{d}x &= \int_\Omega \varphi F^k \cdot H^k \,\mathrm{d}x + \int_\Omega \varphi F^k \cdot G^\infty \,\mathrm{d}x + \int_\Omega \varphi F^k \cdot \nabla w^k \,\mathrm{d}x \\ &= \int_{\mathbb{R}^n} \varphi F^k \cdot H^k \,\mathrm{d}x + \int_{\mathbb{R}^n} \varphi F^k \cdot G^\infty \,\mathrm{d}x + \int_{\mathbb{R}^n} \varphi F^k \cdot \nabla(\psi w^k) \,\mathrm{d}x. \end{aligned} \tag{42}$$

Thanks to (23) and (40), we derive

$$\lim_{k\to\infty}\int_{\mathbb{R}^n}\varphi F^k\cdot H^k\,\mathrm{d}x=0. \tag{43}$$

Moreover, by using again (40) we have

$$\lim_{k\to\infty}\int_{\mathbb{R}^n}\varphi F^k\cdot G^\infty\,\mathrm{d}x=\int_{\mathbb{R}^n}\varphi F^\infty\cdot G^\infty\,\mathrm{d}x=\int_{\Omega}\varphi F^\infty\cdot G^\infty\,\mathrm{d}x. \tag{44}$$

It remains to study the last term in (42). We have that $(\psi w^k)_k\subset W_0^{1,p}(\Omega')$ and

$$\psi w^k\to 0\quad\text{weakly in } W_0^{1,p}(\Omega')\text{ and strongly in } L^{p'}(\Omega'),$$

in view of (22). Therefore, by Proposition 6(b), there exists $(v^k)_k\subset H^{s,p}(\mathbb{R}^n)$ such that

$$\nabla^s v^k=\nabla(\psi w^k)\quad\text{in }\mathbb{R}^n,$$

and by exploiting the estimate (26) we derive

$$v^k\to 0\quad\text{weakly in } H^{s,p'}(\mathbb{R}^n)\text{ and strongly in } L^{p'}(\mathbb{R}^n).$$

Therefore, we can apply Theorem 3 to derive

$$\lim_{k\to\infty}\int_{\mathbb{R}^n}\varphi F^k\cdot\nabla(\psi w^k)\,\mathrm{d}x=\lim_{k\to\infty}\int_{\mathbb{R}^n}\varphi F^k\cdot\nabla^s v^k\,\mathrm{d}x=0. \tag{45}$$

In view of (42)–(45) we finally get (41). □

The last version of the fractional Div-Curl lemma we present is a mixed formulation, stated in the simplified setting of Theorem 2, in which the classical gradient in the product is replaced by a fractional gradient.

Theorem 5 (Second Mixed Local–Nonlocal Div-Curl Lemma) *Let $(F^k)_k\subset L^p_{\mathrm{loc}}(\Omega;\mathbb{R}^n)$, $F^\infty\in L^p_{\mathrm{loc}}(\Omega;\mathbb{R}^n)$, $(w^k)_k\subset H^{s,p'}(\mathbb{R}^n)$, and $w^\infty\in H^{s,p'}(\mathbb{R}^n)$ satisfy as $k\to\infty$*

$$F^k\to F^\infty\qquad\textit{weakly in } L^p_{\mathrm{loc}}(\Omega;\mathbb{R}^n), \tag{46}$$

$$\nabla^s w^k\to\nabla^s w^\infty\qquad\textit{weakly in } L^{p'}_{\mathrm{loc}}(\Omega;\mathbb{R}^n). \tag{47}$$

Assume that as $k\to\infty$

$$\operatorname{div}F^k\to\operatorname{div}F^\infty\quad\textit{strongly in } W^{-1,p}_{\mathrm{loc}}(\Omega). \tag{48}$$

Then, as $k \to \infty$

$$F^k \cdot \nabla^s w^k \to F^\infty \cdot \nabla^s w^\infty \quad \textit{weakly* in } \mathcal{D}'(\Omega). \tag{49}$$

Proof We fix a bounded open set $\Omega' \Subset \Omega$ with smooth boundary.

Let $(u^k)_k \in W^{1,p}_{\mathrm{loc}}(\mathbb{R}^n)$ and $u^\infty \in W^{1,p}_{\mathrm{loc}}(\mathbb{R}^n)$ be the functions given by Proposition 6(a) associated to $(w^k)_k \subset H^{s,p'}(\mathbb{R}^n)$ and $w^\infty \in H^{s,p'}(\mathbb{R}^n)$, respectively. By Poincare's inequality, there exists a constant $C = C(n, p, \Omega') > 0$, independent of $k \in \mathbb{N}$, and a sequence $(c^k)_k \subset \mathbb{R}$ satisfying for all $k \in \mathbb{N}$

$$\|u^k - c^k\|_{L^p(\Omega')} \leq C\|\nabla u^k\|_{L^p(\Omega';\mathbb{R}^n)} = C\|\nabla^s w^k\|_{L^p(\Omega';\mathbb{R}^n)} \leq C.$$

Hence, we can find a subsequence $(k_h)_h \subset \mathbb{N}$ and a function $z^\infty \in W^{1,p}(\Omega')$ such that as $h \to \infty$

$$z^{k_h} := u^{k_h} - c^{k_h} \to z^\infty \quad \text{weakly in } W^{1,p}(\Omega'). \tag{50}$$

By (47), for all $\Psi \in C_c^\infty(\Omega'; \mathbb{R}^n)$ we have

$$\begin{aligned}\int_{\Omega'} \nabla z^\infty \cdot \Psi \, dx &= \lim_{h\to\infty} \int_{\Omega'} \nabla z^{k_h} \cdot \Psi \, dx \\ &= \lim_{h\to\infty} \int_{\Omega'} \nabla u^{k_h} \cdot \Psi \, dx = \lim_{h\to\infty} \int_{\Omega'} \nabla^s w^{k_h} \cdot \Psi \, dx \\ &= \int_{\Omega'} \nabla^s w^\infty \cdot \Psi \, dx.\end{aligned}$$

Hence, $\nabla z^\infty = \nabla^s w^\infty$ in Ω'. Thus in view of (46), (48), and (50) we can apply Theorem 2 and obtain that as $h \to \infty$

$$F^{k_h} \cdot \nabla^s w^{k_h} = F^{k_h} \cdot \nabla z^{k_h} \to F^\infty \cdot \nabla z^\infty = F^\infty \cdot \nabla^s w^\infty \quad \text{weakly* in } \mathcal{D}'(\Omega').$$

Since this holds for every subsequence, we derive (49) in Ω'. By the arbitrariness of $\Omega' \Subset \Omega$, the conclusion follows. □

Remark 10 We point out that these two versions of the mixed local-nonlocal Div-Curl lemma, although partial (since they do not involve the fractional curl), still have useful applications. For instance, although not stated explicitly, such results are used in the proof of the compactness of fractional linear operators with oscillating coefficients in [5, Theorem 3.1].

It is clear that, by exploiting the definition of ∇^s, one can also introduce a fractional curl operator, following Definition 1. Therefore, it would be interesting to extend these fractional Div-Curl lemmas to the case where the second sequence has a strongly converging fractional curl. The main issues are twofold. On the one hand, one must find suitable elliptic regularity estimates in order to derive a fractional Helmholtz decomposition. On the other hand, as explained in Remark 8, the strong convergence (34) is required to hold on the entire space $\mathbb{R}^n$, which is not straightforward to obtain from elliptic regularity results, as these are typically local.

Acknowledgments The author wishes to thank Alessandro Carbotti and Alberto Maione for many helpful discussions on the topic.

Funding Information The author is a member of the Gruppo Nazionale per l'Analisi Matematica, la Probabilità e le loro Applicazioni (INdAM-GNAMPA), and acknowledges the support of the the INdAM-GNAMPA 2025 Project "DISCOVERIES - Difetti e Interfacce in Sistemi Continui: un'Ottica Variazionale in Elasticità con Risultati Innovativi ed Efficaci Sviluppi" (CUP: E5324001950001). The author acknowledges also the support of the Centre de Recerca Matemàtica of Barcelona (CRM), under the International Programme for Research in Groups (IP4RG). Finally, the author has been founded by the European Union-NextGenerationEU under the Italian Ministry of University and Research (MUR) National Centre for HPC, Big Data and Quantum Computing (CN_00000013 - CUP: E13C22001000006).

Competing Interests The author has no conflicts of interest to declare that are relevant to the content of this chapter.

References

1. Alicandro, R., Braides, A., Solci, M., Stefani, G.: Topological singularities arising from fractional-gradient energies. Math. Ann. **393**, 71–111 (2025)
2. Baldi, A., Franchi, B., Tchou, N., Tesi, M.C.: Compensated compactness for differential forms in Carnot groups and applications. Adv. Math. **223**, 1555–1607 (2010)
3. Bellido, J.C., Cueto, J., Mora-Corral, C.: Fractional Piola identity and polyconvexity in fractional spaces. Ann. Inst. H. Poincaré C Anal. Non Linéaire **37**, 955–981 (2020)
4. Briane, M., Casado-Díaz, J, Murat, F.: The div-curl lemma "trente ans après": an extension and an application to the G-convergence of unbounded monotone operators. J. Math. Pures Appl. **91**, 476–494 (2009)
5. Caponi, M., Carbotti, A., Maione, A.: H-compactness for nonlocal linear operators in fractional divergence form. Calc. Var. Partial Differ. Equ. **64**, Paper No. 290 (2025)
6. Comi, G.E., Stefani, G.: A distributional approach to fractional Sobolev spaces and fractional variation: existence of blow-up. J. Funct. Anal. **277**, 3373–3435 (2019)
7. Comi, G.E., Stefani, G.: A distributional approach to fractional Sobolev spaces and fractional variation: asymptotics I. Rev. Mat. Complut. **36**, 491–569 (2023)
8. Conti, S., Dolzmann, G., Müller, S.: The div-curl lemma for sequences whose divergence and curl are compact in $W^{-1,1}$. C. R. Math. Acad. Sci. Paris **349**, 175–178 (2011)
9. Di Fratta, G., Fiorenza, A.: A short proof of local regularity of distributional solutions of Poisson's equation. Proc. Am. Math. Soc. **148**, 2143–2148 (2020)
10. Evans, L.C.: Weak convergence methods for nonlinear partial differential equations. CBMS Regional Conference Series in Mathematics, 74. Published for the Conference Board of the Mathematical Sciences, Washington, DC. American Mathematical Society, Providence (1990)
11. Evans, L.C.: Partial Differential Equations, 2nd edn. Graduate Studies in Mathematics, 19. American Mathematical Society, Providence (2010)
12. Fernández Bonder, J., Ritorto, A., Salort, A.M.: H-convergence result for nonlocal elliptic-type problems via Tartar's method. SIAM J. Math. Anal. **49**, 2387–2408 (2017)

13. Giaquinta, M., Martinazzi, L.: An Introduction to the Regularity Theory for Elliptic Systems, Harmonic Maps and Minimal Graphs, 2nd edn. Appunti. Scuola Normale Superiore di Pisa (Nuova Serie), 11. Edizioni della Normale, Pisa (2012)
14. Kreisbeck, C., Schönberger, H.: Quasiconvexity in the fractional calculus of variations: characterization of lower semicontinuity and relaxation. Nonlinear Anal. **215**, Paper No. 112625 (2022)
15. Maione, A., Paronetto, F., Vecchi, E.: G-convergence of elliptic and parabolic operators depending on vector fields. ESAIM Control Optim. Calc. Var. **29**, Paper No. 8 (2023)
16. Murat, F.: Compacité par compensation. Ann. Scuola Norm. Sup. Pisa Cl. Sci. **5**, 489–507 (1978)
17. Murat, F.: Compacité par compensation: condition nécessaire et suffisante de continuité faible sous une hypothèse de rang constant. Ann. Scuola Norm. Sup. Pisa Cl. Sci. **8**, 69–102 (1981)
18. Shieh, T.-T., Spector, D.E.: On a new class of fractional partial differential equations. Adv. Calc. Var. **8**, 321–336 (2015)
19. Shieh, T.-T., Spector, D.E.: On a new class of fractional partial differential equations II. Adv. Calc. Var. **11**, 289–307 (2018)
20. Šilhavý, M.: Fractional vector analysis based on invariance requirements (critique of coordinate approaches), Contin. Mech. Thermodyn. **32**, 207–228 (2020)
21. Simader, C.G., Sohr, H.: The Dirichlet problem for the Laplacian in bounded and unbounded domains. A new approach to weak, strong and $(2+k)$-solutions in Sobolev-type spaces. Pitman Research Notes in Mathematics Series, 360. Longman, Harlow (1996)
22. Tartar, L.: Une nouvelle méthode de résolution d'équations aux dérivées partielles non linéaires. Journées d'Analyse Non Linéaire (Proc. Conf., Besançon, 1977), pp. 228–241. Lecture Notes in Mathematics, 665. Springer, Berlin (1978)
23. Tartar, L.: Compensated Compactness and Applications to Partial Differential Equations. Nonlinear Analysis and Mechanics: Heriot-Watt Symposium, vol. IV, pp. 136–212. Research Notes in Mathematics, 39. Pitman, Boston, Massachusetts, London (1979)
24. Tartar, L.: The Compensated Compactness Method Applied to Systems of Conservation Laws. Systems of Nonlinear Partial Differential Equations (Oxford, 1982), pp. 263–285. NATO Advanced Science Institutes Series C: Mathematical and Physical Sciences, 111. D. Reidel Publishing Co., Dordrecht (1983)
25. Tartar, L.: The General Theory of Homogenization. A Personalized Introduction. Lecture Notes of the Unione Matematica Italiana, 7. Springer-Verlag, Berlin; UMI, Bologna (2009)

The Pólya-Szegö Principle in the Fractional Setting: A Glimpse on Nonlocal Functional Inequalities

Alessandro Carbotti

Abstract In this chapter we present the fractional Pólya–Szegő principle and its main consequences in the study of nonlocal functional inequalities. In particular, we show how symmetrization methods work also in the fractional setting and yield sharp results such as isoperimetric type inequalities. Further developments including stability issues and generalizations in the anisotropic and the Gaussian setting are also discussed.

1 Introduction

Fractional Sobolev spaces and nonlocal operators have gained significant attention in the last decades. One reason is their natural appearance in various fields, including probability theory, mathematical physics, harmonic analysis, and geometric measure theory. These nonlocal models provide more accurate descriptions of phenomena characterized by long-range interactions or anomalous diffusion, such as in Lévy processes, materials with memory, image processing, and finance, see e.g. [9, 12, 36, 38].

Functional inequalities play a pivotal role in understanding the qualitative and quantitative behaviour of solutions to partial differential equations. They provide essential tools for establishing existence, regularity, stability, and concentration properties of solutions, as well as driving the formulation of variational problems and numerical schemes.

In the local framework functional inequalities have deep geometric content and powerful analytic consequences. Moving to the nonlocal setting many of the aforementioned features are inherited, though additional issues arise owing to long range interactions. In particular, the lack of locality affects the behaviour of nonlocal functionals under rearrangement.

A. Carbotti (✉)
Dipartimento di Matematica e Fisica "Ennio De Giorgi", Università del Salento, Lecce, Italy
e-mail: alessandro.carbotti@unisalento.it

J. Duran i Lamiel, A. Maione (eds.), *New Frontiers in Homogenization and Fractional Calculus*, Trends in Mathematics,
https://doi.org/10.1007/978-3-032-20098-3_3

This chapter focuses on functional inequalities involving geometric quantities enjoying a variational characterization and derived using the *Pólya-Szegö principle* based on the concept of symmetric decreasing rearrangement.

In the book [49] by G. Pólya and G. Szegö, the authors prove that for every $1 \leq p < \infty$ and $u \in W^{1,p}(\mathbb{R}^N, [0, \infty))$ the symmetric decreasing rearrangement u^* (see the following Definition 2) also belongs to $W^{1,p}(\mathbb{R}^N, [0, \infty))$ and

$$\int_{\mathbb{R}^N} |\nabla u^*|^p dx \leq \int_{\mathbb{R}^N} |\nabla u|^p dx. \tag{1}$$

Rearrangement techniques allow one to reduce more involved variational problems to radial or one-dimensional ones, and this reduction may simplify the identification of extremal profiles and optimal constants in functional inequalities.

The rest of the chapter is organized as follows: In Sect. 2 we introduce some notations and preliminary definitions. In Sect. 3 we state and prove in Theorem 2 the Pólya-Szegö principle. In Sect. 4 we recall some of the most famous applications of the Pólya-Szegö inequality, namely

1. The fractional Sobolev inequality in Sect. 4.1
2. The fractional isoperimetric inequality in Sect. 4.2
3. The Faber-Krahn inequality for the fractional p-laplacian in Sect. 4.3
4. A Talenti-type comparison for solutions to fractional equations in Sect. 4.4.

In Sect. 5 we sketch an alternative proof of Theorem 2 in the case $p = 2$ based on the Extension Techniques of [10] and partial symmetrization. In Sect. 6 the problem of stability of the abovementioned inequalities is treated. In Sect. 7 some generalizations involving possibly anisotropic nonlocal functionals are taken into account. The chapter concludes with some final comments and open problems set in Sect. 8.

2 Notations and Preliminaries

Let $s \in (0, 1)$, $N \in \mathbb{N}$ and $1 \leq p < \infty$. The fractional Sobolev space $W^{s,p}(\mathbb{R}^N)$ is defined as the space of functions $u \in L^p(\mathbb{R}^N)$ such that the following quantity

$$[u]_{W^{s,p}(\mathbb{R}^N)} := \left(\int_{\mathbb{R}^N} \int_{\mathbb{R}^N} \frac{|u(x) - u(y)|^p}{|x - y|^{N+sp}} \, dx \, dy \right)^{1/p} \tag{2}$$

is finite. We refer to $[\cdot]_{W^{s,p}(\mathbb{R}^N)}$ as *Gagliardo-Slobodeckij seminorm.*

It is easy to check that the space $W^{s,p}(\mathbb{R}^N)$ endowed with the norm

$$\|u\|_{W^{s,p}(\mathbb{R}^N)} = \left(\|u\|^p_{L^p(\mathbb{R}^N)} + [u]^p_{W^{s,p}(\mathbb{R}^N)} \right)^{1/p} \tag{3}$$

is a Banach space. In particular, when $p = 2$, $W^{s,2}(\mathbb{R}^N)$ with the norm (3) is a Hilbert space which we will indicate in the sequel as $H^s(\mathbb{R}^N)$. It is worth mentioning that the Gagliardo seminorms rescaled respectively by a factor s when $s \to 0^+$ or $1-s$ when $s \to 1^-$, approach in the limit the $L^p(\mathbb{R}^N)$ norm or the $W^{1,p}(\mathbb{R}^N)$ seminorm, both in the pointwise and in the Γ-convergence sense see [4, 22, 45, 50]. We refer to [26] for further details and properties.

2.1 The Fractional Laplacian

The prototype of nonlocal fractional operator is the fractional laplacian $(-\Delta)^s$ which is defined as the Fourier multiplier with symbol $|\cdot|^{2s}$, namely

$$(-\Delta)^s u(x) = \mathcal{F}^{-1}\left(|\cdot|^{2s}\mathcal{F}(u)\right)(x), \tag{4}$$

where $\mathcal{F}$ denotes the Fourier transform on $L^1(\mathbb{R}^N) \cap L^2(\mathbb{R}^N)$

$$\mathcal{F}(u)(\xi) = \int_{\mathbb{R}^N} u(x)e^{-ix\cdot\xi}\,dx.$$

The fractional laplacian enjoys many equivalent definitions on the whole of $\mathbb{R}^N$. We give here three among others which will be useful in the sequel

1. **Fractional laplacian as integro-differential operator**
For $u \in H^s(\mathbb{R}^N)$, the fractional laplacian $(-\Delta)^s u(x)$ is given by

$$(-\Delta)^s u(x) = C_{N,s}\,\text{P.V.}\int_{\mathbb{R}^N} \frac{u(x)-u(y)}{|x-y|^{N+2s}}\,dy, \tag{5}$$

where P.V. stands for the Cauchy principal value. Moreover, if $u \in H^s(\mathbb{R}^N) \cap C^2(B_R)$ for some $R > 0$, the principal value may be dropped and the integral has to be intended in the Lebesgue sense.

Remark 1 The assumption $u \in H^s(\mathbb{R}^N) \cap C^2(B_R)$ can be relaxed by requiring that $u \in C^{0,2s+\varepsilon}(B_R)$ if $0 < s < \frac{1}{2}$ or $u \in C^{1,2s-1+\varepsilon}(B_R)$ for some $\varepsilon > 0$ if $\frac{1}{2} \leq s < 1$ and that $\frac{u}{1+|\cdot|^{N+2s}} \in L^1(\mathbb{R}^N)$ so that $(-\Delta)^s u(x)$ is still pointwise well-defined almost everywhere.

2. **Fractional laplacian via heat flow**
Given a sufficiently smooth function u with a certain growth at infinity, the *heat semigroup* $e^{t\Delta}$ applied to u is given by $U(t,x) = e^{t\Delta}u := G_t * u$ where, for every $t > 0$, G_t denotes the *heat kernel*

$$G_t(\cdot) := (4\pi t)^{-N/2}e^{-\frac{|\cdot|^2}{4t}}. \tag{6}$$

It can be proved that U solves the Cauchy problem

$$\begin{cases} U_t = \Delta U & \text{in} \quad (0,\infty)\times\mathbb{R}^N \\ U(0,x) = u(x) & \text{in} \quad \mathbb{R}^N. \end{cases}$$

Since for every $\lambda > 0$ and $s \in (0,1)$ the following identity holds true

$$\lambda^s = \frac{1}{\Gamma(-s)}\int_0^\infty \frac{e^{-t\lambda}-1}{t^{s+1}}dt,$$

where Γ denotes the Euler's Gamma function, by means of functional calculus one can replace the positive number λ with the positive operator $-\Delta$ and consider the fractional powers $(-\Delta)^s$ through the *Bochner Subordination formula*

$$(-\Delta)^s u = \frac{1}{\Gamma(-s)}\int_0^\infty \frac{e^{t\Delta}u - u}{t^{s+1}}dt. \tag{7}$$

We refer e.g. to the book [43] for further details on the definition of fractional powers of operators via functional calculus.

3. **Fractional laplacian via "Harmonic" extension**
   The fractional laplacian enjoys the following characterization in terms of the Dirichlet-to-Neumann map associated to a divergence form local operator defined on the upper half-space $\mathbb{R}^{N+1}_+$ with degenerate or singular coefficients. Namely

Theorem 1 ([10, Section 3]) *Let $u : \mathbb{R}^N \to \mathbb{R}$ and $U : \mathbb{R}^{N+1}_+ \to \mathbb{R}$ be the unique weak solution of the problem*

$$\begin{cases} div(y^{1-2s}\nabla U) = 0 & in \quad \mathbb{R}^{N+1}_+ \\ U(x,0) = u(x) & in \quad \mathbb{R}^N. \end{cases} \tag{8}$$

Then

$$(-\Delta)^s u(x) = -c_s \lim_{y\to 0^+} y^{1-2s}\partial_y U(x,y),$$

and

$$[u]^2_{H^s(\mathbb{R}^N)} = c_s \inf_{\substack{U\in H^1_{\mathrm{loc}}(\mathbb{R}^{N+1}_+)\\ U(\cdot,0)=u}} \int_0^\infty y^{1-2s}dy \int_{\mathbb{R}^N}\left(|\nabla_x U|^2 + |\partial_y U|^2\right)dx \tag{9}$$

Remark 2 In the case $s = \frac{1}{2}$, Problem (8) is the harmonic extension on the upper half-space, and the "half"-laplacian $(-\Delta)^{1/2}$ is the inward normal derivative of the extension U at $\{y = 0\}$.

Remark 3 The presence of the constant $C_{N,s} := \frac{4^s\Gamma\left(\frac{N}{2}+s\right)}{\pi^{N/2}\Gamma(1-s)}$ in (5) makes consistent the equivalence among the several definition of $(-\Delta)^s$. See also [9, Section 3].

Remark 4 It is worth noticing that, for $1 < p < \infty$, one can also consider a nonlinear counterpart of $(-\Delta)^s$ given by the *fractional p-laplacian*

$$(-\Delta)_p^s u(x) := \text{P.V.} \int_{\mathbb{R}^N} \frac{|u(x) - u(y)|^{p-2}(u(x) - u(y))}{|x - y|^{N+sp}} dy.$$

We refer to [24] for analogous characterization of $(-\Delta)_p^s$ as in the rest of this subsection.

Definition 1 (Schwarz Symmetrization) We define the symmetric decreasing rearrangement (or *Schwarz symmetral*) Ω^* of an open and bounded set Ω as the ball centered at the origin with the same measure of Ω, namely

$$\Omega^* = \{x \in \mathbb{R}^N : \ \omega_N |x|^N < |\Omega|\},$$

where $|\cdot|$ stands as usual for the N-dimensional Lebesgue measure of a measurable set and ω_N denotes the volume of the unit ball in $\mathbb{R}^N$.

Definition 2 (Symmetric Decreasing Rearrangement of a Function) Let $u : \mathbb{R}^N \to [0, \infty)$ be a measurable function. We say that u *vanishes at infinity* if

$$|\{u(x) > t\}| < \infty \quad \text{for every} \quad t > 0.$$

The *symmetric decreasing rearrangement* $u^* : \mathbb{R}^N \to [0, \infty)$ of a function u vanishing at infinity is defined through the *layer-cake* representation formula as

$$u^*(x) = \int_0^\infty \chi_{\{f(x)>t\}^*} dt.$$

The *distribution function* of u is defined by

$$\mu_u(t) := |\{x \in \mathbb{R}^N : u(x) > t\}|, \quad t > 0,$$

and u^* is the unique radially symmetric decreasing function which is *equimeasurable* with respect u, i.e.

$$\mu_{u^*}(t) = \mu_u(t), \quad \text{for all } t > 0.$$

More explicitly, there exists a function $u_0 : [0, \infty) \to [0, \infty)$, decreasing and right-continuous, such that $u^*(x) = u_0(\omega_N |x|^N)$ and, for every $r > 0$

$$u_0(r) = \sup\{t \geq 0 : \mu_u(t) > r\}.$$

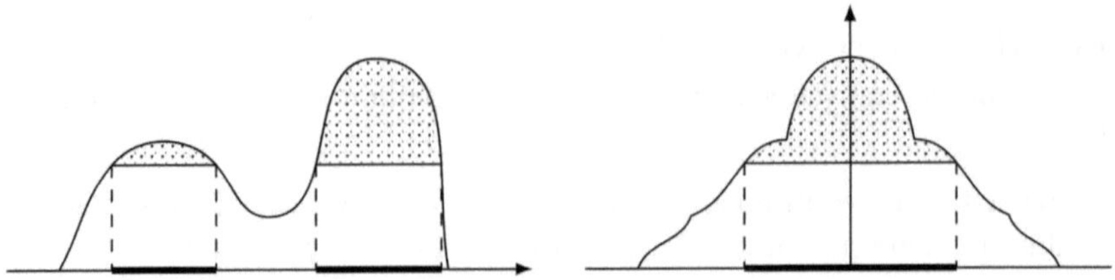

Fig. 1 Schwarz rearrangement of a 1D function (Image kindly provided by Andrea Gentile)

See the following Fig. 1 for a graphical comparison between a positive function and its symmetric decreasing rearrangement.

Remark 5 Notice that Ω^* is the unique set such that $\chi_{\Omega^*} = (\chi_\Omega)^*$.

Two remarkable results involving symmetric decreasing rearrangement are the following

Lemma 1 (Rearrangement Preserves L^p Norms) *Let $1 \leq p \leq \infty$. For every non-negative $u \in L^p(\mathbb{R}^N)$ it holds that $u^* \in L^p(\mathbb{R}^N)$ with*

$$\left\|u^*\right\|_{L^p(\mathbb{R}^N)} = \|u\|_{L^p(\mathbb{R}^N)} .$$

Lemma 2 (Riesz Rearrangement Inequality) *Let $f, g, h : \mathbb{R}^N \to [0, \infty)$ measurable functions. Then*

$$\int_{\mathbb{R}^N} f(x)g(x-y)h(y)dxdy \leq \int_{\mathbb{R}^N} f^*(x)g^*(x-y)h^*(y)dxdy.$$

The Schwarz symmetrization is not the only rearrangement that allows for the Pólya-Szegö inequality (1). We introduce the *Steiner symmetrization* of a set

Definition 3 (Steiner Symmetrization) Let $\Omega \subset \mathbb{R}^N$ be a measurable set and let $e \in \mathbb{S}^{N-1}$ be a fixed direction. For each $x \in e^\perp$, consider the one-dimensional section

$$\Omega_x := \{t \in \mathbb{R} : x + te \in \Omega\}.$$

The *Steiner symmetral* of Ω with respect to the direction e is the set

$$\Omega^\sharp := \{x + te : x \in e^\perp,\ |t| < \tfrac{1}{2}\mathcal{L}^1(\Omega_x)\},$$

where $\mathcal{L}^1(\Omega_x)$ denotes the one-dimensional Lebesgue measure of Ω_x.

In other words, each section Ω_x along the line parallel to e is replaced by the centered interval of the same length, symmetric with respect to the hyperplane $e^\perp$.

Definition 4 (Steiner Symmetrization of a Function) Let $f : \mathbb{R}^N \to [0, \infty)$ be a measurable function and let $e \in \mathbb{S}^{N-1}$ be a fixed direction. For each $x \in e^{\perp}$, consider the one-dimensional function

$$f_x(t) := f(x + te), \qquad t \in \mathbb{R}.$$

We denote by $f_x^{\sharp}$ the one-dimensional symmetric decreasing rearrangement of f_x, i.e., the unique nonnegative, even, and nonincreasing function such that

$$\mathcal{L}^1 (\{t \in \mathbb{R} : f_x(t) > \lambda\}) = \mathcal{L}^1 \left(\{t \in \mathbb{R} : f_x^{\sharp}(t) > \lambda\}\right) \quad \text{for all } \lambda > 0.$$

The *Steiner symmetrization* of f with respect to the direction e is then defined by

$$f^{\sharp}(x + te) := f_x^{\sharp}(t), \qquad x \in e^{\perp},\ t \in \mathbb{R}.$$

In other words, along each line parallel to e, the function is replaced by its one-dimensional symmetric decreasing rearrangement centered with respect to the hyperplane $e^{\perp}$.

Remark 6 We highlight in Table 1 some analogies and differences between Steiner and Schwarz symmetrizations. Both the first and the second one aim to replace a given set or function with a more symmetric one, preserving the volume measure or

Table 1 Analogies and differences between Schwarz and Steiner symmetrization

	Steiner symmetrization	Schwarz symmetrization
For sets	For each line parallel to a fixed direction, replace the 1D section of E by an interval centered on the hyperplane with the same measure	Replace each section of E by a ball centered at the origin having the same measure
For functions	For each line parallel to a fixed direction, replace the 1D level sets of u by an interval centered on the hyperplane with the same measure	Replace each level set of u by a ball centered at the origin having the same measure
Symmetry of rearranged functions	1D even symmetry with respect to a chosen direction	Radially symmetric
Symmetry of rearranged sets	Axial symmetry with respect to a hyperplane	Radial symmetry
Iterative use	Repeated Steiner symmetrizations in many directions approximate Schwarz symmetrization	Fully symmetric after one step
Effect on convexity	Preserves convexity: if E is convex, then its Steiner symmetral is convex	Does not necessarily preserve convexity (except for balls)
Typical applications	Useful in anisotropic problems, or when convexity must be preserved	Comparison between differential and variational problems with the radial symmetric counterpart

the L^p-norm and proving the monotonicity of Dirichlet energies such as in (10) or for convolution-type integrals such as in Lemma 2.

3 The Pólya-Szegö Principle for Gagliardo Fractional Seminorms

We are now in the position to state the main result of this chapter

Theorem 2 (Fractional Pólya–Szegö Inequality) *Let* $1 \leq p < \infty$ *and* $u \in W^{s,p}(\mathbb{R}^N)$. *Then* $u^* \in W^{s,p}(\mathbb{R}^N)$ *and*

$$[u^*]_{W^{s,p}(\mathbb{R}^N)} \leq [u]_{W^{s,p}(\mathbb{R}^N)}. \tag{10}$$

If $p = 1$ *then equality holds if and only if* u *does not change sign and the level set* $\{u > \tau\}$ *is a ball for almost every* $\tau > 0$. *If* $p > 1$ *equality holds if and only if* u *is the translate of a symmetric decreasing function.*

Proof Since $||u(x)| - |u(y)|| \leq |u(x) - u(y)|$ and equality for every $x, y \in \mathbb{R}^N$ holds if and only if $u \geq 0$ we reduce ourselves to prove the claim for non-negative functions. Moreover, we assume for the moment that u is also bounded. Let $G_t(\cdot)$ be the heat kernel (6). The fractional seminorm (2) can be represented (up to a multiplicative constant) as:

$$\begin{aligned} [u]^p_{W^{s,p}(\mathbb{R}^N)} &= C_{N,s,p} \int_0^\infty t^{-1-\frac{sp}{2}} \int_{\mathbb{R}^N} \int_{\mathbb{R}^N} |u(x) - u(y)|^p G_t(x-y)\, dx dy\, dt \\ &= C_{N,s,p} \int_0^\infty I_t[u] t^{\frac{N+sp}{2}-1} dt, \end{aligned} \tag{11}$$

where, for every $t > 0$ we have set

$$I_t[u] := \int_{\mathbb{R}^N} \int_{\mathbb{R}^N} |u(x) - u(y)|^p e^{-t|x-y|^2} dx dy.$$

Equality (11) is obtained via Bochner subordination formula (7) since

$$\frac{C_{N,s,p}}{|x-y|^{N+sp}} = \int_0^\infty G_t(x-y) t^{-1-\frac{sp}{2}} dt.$$

We now prove that, for every $t > 0$ rearrangement decreases the functional $I_t[u]$. We write $I_t[u] = I_t^+[u] + I_t^-[u]$, where

$$I_t^\pm[u] := C_{N,s,p} \int_{\mathbb{R}^N} \int_{\mathbb{R}^N} (u(x) - u(y))_\pm^p e^{-t|x-y|^2} dx dy,$$

where, as usual

$$(r)_+ := \begin{cases} r & \text{if} \quad r \geq 0 \\ 0 & \text{if} \quad r < 0 \end{cases}$$

and $(r)_- = (-r)_+$. We firstly argue on $I_t^+[u]$. Since the function $\mathbb{R} \ni z \mapsto J(z) := |z|^p$ is convex, so does J_+ and it has a non-negative and non-decreasing right derivative J'_+, so that we can write

$$(u(x) - u(y))_+^p = \int_0^\infty J'_+(u(x) - \tau)\chi_{\{u \leq \tau\}}(y)d\tau,$$

that yields

$$I_t^+[u] = C_{N,s,p} \int_0^\infty \left(\int_{\mathbb{R}^N} \int_{\mathbb{R}^N} J'_+(u(x) - \tau)e^{-t|x-y|^2} \chi_{\{u \leq \tau\}}(y)dxdy \right) d\tau. \tag{12}$$

Since u is bounded and $|\{u > \tau\}| < \infty$ for every $\tau > 0$ we have that

$$\int_{\mathbb{R}^N} J'_+(u(x) - \tau)dx < \infty,$$

Therefore, denoting by $e_{t,\tau}^+[u]$ the inner integral in (12) and using that $\chi_{\{u \leq \tau\}} = 1 - \chi_{\{u > \tau\}}$, we write

$$\begin{aligned} e_{t,\tau}^+[u] = \left\| e^{-t|\cdot|^2} \right\|_{L^1(\mathbb{R}^N)} \int_{\mathbb{R}^N} J'_+(u(x) - \tau)dx \\ - \int_{\mathbb{R}^N} \int_{\mathbb{R}^N} J'_+(u(x) - \tau)e^{-t|x-y|^2} \chi_{\{u > \tau\}}(y)dxdy. \end{aligned} \tag{13}$$

We notice that the rearrangement leaves invariant the first term in the right-hand side of (13) since $J'_+((u(x) - \tau))^* = J'_+(u^*(x) - \tau)$, while the second term does not decrease thanks to Lemma 2. This proves that $e_{t,\tau}^+[u] \geq e_{t,\tau}^+[u^*]$ for almost every $\tau > 0$ and for every $t > 0$. Integrating in $\tau \in (0, \infty)$ we conclude that

$$I_t^+[u] \geq I_t^+[u^*]$$

for every $t > 0$. Arguing in the same way for $I_t^-[u]$ exchanging the roles of x and y in the definition of $e_{t,\tau}^+[u]$ and replacing $(\cdot)_+$ with $(\cdot)_-$, we conclude that

$$I_t[u] \geq I_t[u^*] \tag{14}$$

for every $t > 0$.

Multiplying by $t^{\frac{N+sp}{2}-1}$, integrating in $t \in (0, \infty)$ and raising to the power $1/p$ both sides of (14) we obtain:

$$[u^*]_{W^{s,p}(\mathbb{R}^N)} \le [u]_{W^{s,p}(\mathbb{R}^N)},$$

which concludes the first part of the proof.

Now, we characterize the case of equality in (10).

Assume that $I_t^{\pm}[u^*] = I_t^{\pm}[u]$ for some $u \in W^{s,p}(\mathbb{R}^N) \cap L^\infty(\mathbb{R}^N)$ and every $t > 0$. By (13) we have that $e_{t,\tau}^{\pm}[u^*] = e_{t,\tau}^{\pm}[u]$ for every $t > 0$ and almost every $\tau > 0$. By Lieb's strict rearrangement inequality [40, Lemma 3], we have that for a.e. $\tau > 0$ there exists $a_\tau \in \mathbb{R}^N$ such that $\chi_{\{u<\tau\}}(x) = \chi_{\{u^*<\tau\}}(x - a_\tau)$ and

$$J'_\pm(u(x) - \tau) = J'_\pm(u^*(x - a_\tau) - \tau).$$

Now, if $p = 1$ then $J_+(t) = t_+$ for every $t \in \mathbb{R}$ and this means that the superlevel sets $\{u > \tau\}$ are balls for almost every $\tau > 0$. Otherwise if $p > 1$ then J_+ is strictly convex in $(0, \infty)$ and so J'_+ is strictly increasing in $(0, \infty)$ which yields

$$u(x) = u^*(x - a_\tau) \tag{15}$$

for almost every $x \in \mathbb{R}^N$, $\tau > 0$. Since the left-hand side in (15) does not depend on τ we have that u is a translate of a symmetric decreasing function, and this concludes the proof for $u \in L^\infty(\mathbb{R}^N)$.

Now we remove the assumption that u is bounded, that is, we claim that (14) holds for any non-negative u with $I_t[u]$ and $|\{u > \tau\}|$ finite for almost every $\tau > 0$ and every $t > 0$. To see this, replace u by $u_M = \min\{u, M\}$ and note that $(u_M)^* = (u^*)_M =: u_M^*$ and $I_t[u_M] \le I_t[u]$. By monotone convergence the claim follows easily from the inequality

$$I_t[u_M] \ge I_t[u_M^*]. \tag{16}$$

Finally, we characterize the cases of equality for general u. Assume that $I_t^+[u] = I_t^+[u^*]$, for any $M > 0$ we decompose

$$u = u_M + v_M, \qquad u_M := \min\{u, M\},$$

and find

$$I_t^+[u] = I_t^+[u_M] + I_t^+[v_M] + \iint_{\mathbb{R}^N\times\mathbb{R}^N} F_M\big(v_M(x), u_M(y)\big) e^{-t|x-y|^2}\, dx\, dy \tag{17}$$

with

$$F_M(v, u) := J_+(v + M - u) - J_+(v) - J_+(M - u).$$

Note that since J_+ is convex with $J_+(0) = 0$, one has $F_M(v, u) \ge 0$ for $0 \le u \le M$ and $v \ge 0$. Hence all three terms on the right-hand side of (17) are non-negative and finite. Note that replacing u by u^* amounts to replacing u_M and v_M by u_M^*

and v_M^*, respectively. Below we shall prove that the double integral in (17) does not increase if both u_M and v_M are replaced by u_M^* and v_M^*. Moreover, by (16), $I_t^+[v_M] \geq I_t^+[v_M^*]$. Hence if $I_t^+[u] = I_t^+[u^*]$, then $I_t^+[u_M] = I_t^+[u_M^*]$ for all $M > 0$. Using (15) one easily concludes that u is the translate of a symmetric decreasing function.

It suffices to prove that the double integral in (17) does not increase under rearrangement. Since J_+' is increasing, we have $J_+'(t) = \int_0^t d\mu(\tau)$ for a non-negative measure μ. Hence

$$J_+(t) = \int_0^\infty (t-\tau)_+ \, d\mu(\tau) \quad \text{and} \quad F_M(v,u) = \int_0^\infty f_{M,\tau}(v,u)\, d\mu(\tau),$$

where

$$f_{M,\tau}(v,u) := (v + M - u - \tau)_+ - (v-\tau)_+ - (M-u-\tau)_+.$$

Since the integrand is non-negative for $0 \leq u \leq M$ and $v \geq 0$, it suffices to prove that for almost every τ the double integral

$$\iint_{\mathbb{R}^N \times \mathbb{R}^N} f_{M,\tau}\big(v_M(x), u_M(y)\big) e^{-t|x-y|^2}\, dx\, dy$$

does not increase under rearrangement. We decompose further $f_{M,\tau} = f_{M,\tau}^{(1)} - f_{M,\tau}^{(2)}$ where

$$f_{M,\tau}^{(1)}(v) := v - (v-\tau)_+$$

and

$$f_{M,\tau}^{(2)}(v,u) := v - (v + M - u - \tau)_+ + (M-u-\tau)_+ = \min\{\, v,\ (u - M + \tau)_+ \,\}.$$

Since $f_{M,\tau}^{(1)}$ is bounded and the support of v_M has finite measure, the integral

$$\iint_{\mathbb{R}^N \times \mathbb{R}^N} f_{M,\tau}^{(1)}\big(v_M(x)\big)\, e^{-t|x-y|^2}\, dx\, dy = \left\| e^{-t|\cdot|^2} \right\|_{L^1(\mathbb{R}^N)} \int_{\mathbb{R}^N} f_{M,\tau}^{(1)}\big(v_M(x)\big)\, dx$$

is finite and invariant under rearrangement of v_M for every $t > 0$. Finally, by Fubini Theorem we can write

$$\begin{aligned} &\iint_{\mathbb{R}^N \times \mathbb{R}^N} f_{M,\tau}^{(2)}\big(v_M(x), u_M(y)\big) e^{-t|x-y|^2}\, dx\, dy = \\ &\int_0^\infty \left(\iint_{\mathbb{R}^N \times \mathbb{R}^N} \chi_{\{v_M > \sigma\}}(x)\, e^{-t|x-y|^2}\, \chi_{\{(u_M - M + \tau)_+ > \sigma\}}(y)\, dx\, dy \right) d\sigma. \end{aligned} \tag{18}$$

By Lemma 2, the right-hand side in (18) does not decrease under rearrangement for almost every $\tau > 0$ and for every $t > 0$, which completes the proof. □

Remark 7 In [33, Theorem 1], the fractional Pólya-Szegö inequality is set for $1 \leq p < \frac{N}{s}$ in the homogeneous fractional Sobolev space $\dot{W}^{s,p}(\mathbb{R}^N) := \overline{C_c^\infty(\mathbb{R}^N)}^{[\cdot]_{W^{s,p}(\mathbb{R}^N)}} = \left\{ u \in L^{\frac{Np}{N-sp}}(\mathbb{R}^N);\ [u]_{W^{s,p}(\mathbb{R}^N)} < \infty \right\}$.

Inequality (14) is a particular case of a more general result proved in [33, Lemma A.2].

Lemma 3 *Let $J : \mathbb{R} \to [0, \infty)$ be a convex function with $J(0) = 0$, $k \in L^1(\mathbb{R}^N)$ be a symmetric decreasing function and*

$$E[u] := \int_{\mathbb{R}^N} \int_{\mathbb{R}^N} J(u(x) - u(y))K(x - y)dxdy.$$

Then, for every non-negative and measurable u with $E[u]$ and $|\{u > \tau\}|$ finite for every $\tau > 0$ one has that

$$E[u] \geq E[u^*]. \tag{19}$$

Remark 8 Inequality (19) has been proved also in [1, Theorem 9.2 (i)] under the additional hypoteses of even symmetry of J and finiteness of the quantity $\int_{\mathbb{R}^N} J(u(x))dx$. Clearly these hypoteses encompass also the $W^{s,p}(\mathbb{R}^N)$ seminorms thanks to the representation formula (11). The novelty of the proof given in [33] is the *rigidity*, namely, the characterization of the cases of equality in (19).

4 Some Applications

4.1 Fractional Sobolev Inequality

Let $1 \leq p < \frac{N}{s}$. The *fractional Sobolev inequality* states that the continuous embedding

$$W^{s,p}(\mathbb{R}^N) \hookrightarrow L^q(\mathbb{R}^N) \tag{20}$$

holds true for every $q \in [p, p_s^*]$, where $p_s^* := \frac{Np}{N-sp}$ is the *fractional Sobolev critical exponent*. In particular, in the critical case $q = p_s^*$ one can search for a maximizer of the Sobolev quotient

$$S_{s,p}(u) := \frac{\|u\|_{L^{p_s^*}(\mathbb{R}^N)}}{[u]_{W^{s,p}(\mathbb{R}^N)}} \tag{21}$$

among all $u \in W^{s,p}(\mathbb{R}^N) \setminus \{0\}$. Through the Pólya-Szegö inequality one can reduce to the one-dimensional case searching for a radial maximizer to (21). In particular, when $p = 2$, the unique optimal profile up to translations and dilations is explicit and it is given by

$$u_{\max,2}(x) := (1 + |x|^2)^{-\frac{N-2s}{2}},$$

see [21, Theorem 1.1], while for $p \neq 2$ it is only known that the optimal profile $u_{\max,p}$ is radially decreasing with $u_{\max,p}(x) \sim |x|^{-\frac{N-sp}{p-1}}$ as $|x| \to \infty$ (see [6, Theorem 1.1.]). It is worth noticing that for $s \to 1^-$ we recover the behaviour of the maximizers of the classical Sobolev quotient $\mathcal{S}_{1,p}(u)$ commonly known as *Aubin-Talenti bubbles*.

4.2 Fractional Isoperimetric Inequality

Nonlocal perimeters either of fractional type or depending on more general kernels have been widely studied in the last years, since they are related to nonlocal minimal surfaces [11, 44], phase transitions [52], fractal sets [41] and many other problems. The relevant isoperimetric inequalities of qualitative and quantitative type have been proved in [17, 33] and [31, 35], respectively. See also [20] where the authors introduce a notion of fractional perimeter using a distributional approach and [25] where an isoperimetric problem with the competition of two fractional perimeters of different order is studied.

The s-fractional perimeter $P_s(E)$ of a measurable set $E \subset \mathbb{R}^N$ is defined as:

$$P_s(E) = \frac{[\chi_E]_{W^{s,1}(\mathbb{R}^N)}}{2} = \int_E \int_{E^c} \frac{dxdy}{|x-y|^{N+s}}, \tag{22}$$

in particular, E is a set of finite s-fractional perimeter if and only if $\chi_E \in W^{s,1}(\mathbb{R}^N)$.

The fractional Pólya-Szegö principle yields the following

Theorem 3 (Fractional Isoperimetric Inequality) *Among all open and bounded sets, balls minimize the scale invariant isoperimetric ratio, i.e.*

$$\frac{P_s(\Omega)}{|\Omega|^{\frac{N-s}{N}}} \geq \frac{P_s(B)}{|B|^{\frac{N-s}{N}}}. \tag{23}$$

In particular, if $|\Omega| = |B|$ *then*

$$P_s(\Omega) \geq P_s(B),$$

and equality holds if and only if Ω *is a ball.*

Proof It is sufficient to apply Theorem 2 with $u = \chi_\Omega$ and using that $\chi_\Omega^* = \chi_{\Omega^*}$.

□

Remark 9 Inequality (23) can be restated by noticing that $P_s(B) = C|B|^{\frac{N-s}{N}} = C|\Omega|^{\frac{N-s}{N}}$ for some explicit $C = C(N,s)$, (see [37, Proposition 1.1.] replacing s

with $\frac{s}{2}$). In particular, this yields that for every open set Ω with finite measure it holds

$$P_s(\Omega) \geq C|\Omega|^{\frac{N-s}{N}}.$$

This is also a remarkable application of the fractional Sobolev inequality (20) to χ_Ω.

Remark 10 Since $|\chi_E(x) - \chi_E(y)| = |\chi_E(x) - \chi_E(y)|^p$ for every $x, y \in \mathbb{R}^N$ and $1 \leq p < \infty$ it is easy to check that

$$P_s(E) = \frac{1}{2}[\chi_E]^p_{W^{\frac{s}{p},p}(\mathbb{R}^N)}.$$

4.3 *Fractional Dirichlet Eigenvalues and Faber-Krahn Inequality*

Let $1 < p < \frac{N}{s}$ and Ω be a bounded open set in $\mathbb{R}^N$. The Dirichlet eigenvalue problem for $(-\Delta)^s_p$ consists in finding the minimum value λ such that the external value problem

$$\begin{cases} (-\Delta)^s_p u = \lambda|u|^{p-2}u & \text{in} \quad \Omega \\ u = 0 \quad \text{in} \quad \mathbb{R}^N \setminus \Omega \end{cases} \tag{24}$$

admits a nonzero weak solution.

Such a minimum admits a variational characterization in terms of the *Rayleigh quotient*. We define

$$W^{s,p}_0(\Omega) := \overline{C^\infty_c(\Omega)}^{\|\cdot\|_{W^{s,p}(\mathbb{R}^N)}} \subsetneq W^{s,p}(\mathbb{R}^N)$$

and $\mathcal{R}_{\Omega,s,p} : W^{s,p}_0(\Omega) \to [0, \infty)$ defined as

$$\mathcal{R}_{\Omega,s,p}(u) := \frac{[u]^p_{W^{s,p}(\mathbb{R}^N)}}{\|u\|^p_{L^p(\mathbb{R}^N)}} \tag{25}$$

The first Dirichlet eigenvalue for $(-\Delta)^s_p$ is therefore characterized in the following way

$$\lambda^{s,p}_1(\Omega) = \inf_{u \in W^{s,p}_0(\Omega)\setminus\{0\}} \mathcal{R}_{\Omega,s,p}(u). \tag{26}$$

Remark 11 Using Direct Methods of the calculus of variations it can be proved that the infimum is actually a minimum and it is attained in $u = u_\Omega$ where u_Ω is the first Dirichlet eigenfunction, i.e., the unique weak solution of (24).

Remark 12 Notice that the right-hand side in (25) actually depends on Ω. Indeed, since $u = 0$ in $\mathbb{R}^N \setminus \Omega$ we have that

$$\mathcal{R}_{\Omega,s}(u) = \frac{\mathcal{I}_{\Omega,s,p}(u)}{\|u\|^p_{L^p(\Omega)}},$$

where

$$\mathcal{I}_{\Omega,s,p}(u) := \iint_{Q_\Omega} \frac{|u(x) - u(y)|^p}{|x - y|^{N+sp}} dxdy$$

and $Q_\Omega := (\Omega \times \Omega) \cup (\Omega \times \Omega^c) \cup (\Omega^c \times \Omega)$.

The fractional Faber-Krahn inequality goes as follows, see [5, Theorem 3.5]:

Theorem 4 (Fractional Faber–Krahn Inequality) *Among all open and bounded sets balls minimize the scale invariant functional:*

$$|\Omega|^{\frac{sp}{N}} \lambda_1^{s,p}(\Omega) \geq |B|^{\frac{sp}{N}} \lambda_1^{s,p}(B). \tag{27}$$

In particular, if $|B| = |\Omega|$ *we have the Faber-Krahn inequality for the first Dirichlet eigenvalue of* $(-\Delta)^s_p$

$$\lambda_1^{s,p}(\Omega) \geq \lambda_1^{s,p}(B),$$

and equality holds if and only if Ω *is a ball.*

Proof The proof follows by Theorem 2 by using that $\lambda_1^{s,p}(\Omega) = \mathcal{R}_{\Omega,s,p}(u_\Omega) = [u_\Omega]^p_{W^{s,p}(\mathbb{R}^N)}$, where u_Ω is the first Dirichlet eigenfunction with $L^p(\Omega)$ unit norm, and that $u^*_\Omega = u_{\Omega^*}$. □

4.4 *Talenti-Type Comparison for Solutions to Fractional Equations*

The Pólya-Szegö principle allows also in the fractional case for a comparison between a solution to a fractional equation and a solution to a corresponding symmetrized problem. For the sake of concreteness we state the Theorem in the model case of the fractional laplacian. We refer to [30, Theorems 1.1, 1.2] where a broad class of fractional elliptic and parabolic operator is also treated.

Theorem 5 *Let $\Omega \subset \mathbb{R}^N$ be an open and bounded set and $f \in L^2(\Omega)$, $f \geq 0$. If u and v are respectively solutions to*

$$\begin{cases} (-\Delta)^s u = f & in \quad \Omega \\ u = 0 & in \quad \mathbb{R}^N \setminus \Omega \end{cases}$$

and

$$\begin{cases} (-\Delta)^s v = f^* & in \quad \Omega^* \\ v = 0 & in \quad \mathbb{R}^N \setminus \Omega^*, \end{cases}$$

Then

$$\int_{B_r} u^*(x)dx \leq \int_{B_r} v(x)dx \quad \forall r > 0$$

and

$$[u]_{H^s(\mathbb{R}^N)} \leq [v]_{H^s(\mathbb{R}^N)}.$$

5 Pólya-Szegö Inequalities via Partial Symmetrization of Extension Functionals

We now show in the special case $p = 2$ an alternative proof of Theorem 2 based on the characterization of the H^s-seminorm given by Theorem 1 and on the *Partial Symmetrization* of the weighted Dirichlet integral in (9). We give here a brief sketch of the proof

1. We set

$$I_1[U] := \int_0^\infty y^{1-2s} \left(\int_{\mathbb{R}^N} |\nabla_x U(x, y)|^2 dx \right) dy \tag{28}$$

and

$$I_2[U] := \int_0^\infty y^{1-2s} \left(\int_{\mathbb{R}^N} |\partial_y U(x, y)|^2 dx \right) dy \tag{29}$$

where U solves (8).

2. With a slight abuse of notation we indicate again with U^* the Schwarz symmetrization of U with respect the x variable. Therefore the classical Pólya-Szegö inequality (1) yields that

$$\int_{\mathbb{R}^N} |\nabla_x U^*(x, y)|^2 dx \leq \int_{\mathbb{R}^N} |\nabla_x U(x, y)|^2 dx$$

for every $y > 0$. Multiplying by y^{1-2s} and integrating in $(0, \infty)$ we obtain that $I_1[U^*] \leq I_1[U]$.
3. The inequality

$$\int_{\mathbb{R}^N} |\partial_y U^*(x, y)|^2 dx \leq \int_{\mathbb{R}^N} |\partial_y U(x, y)|^2 dx \tag{30}$$

is a consequence of [8, Theorem 1]. Again, by multiplying by y^{1-2s} and integrating in $(0, \infty)$ we obtain that $I_2[U^*] \leq I_2[U]$.
4. Let $V : \mathbb{R}^{N+1}_+ \to [0, \infty)$ the unique weak solution of

$$\begin{cases} \operatorname{div}(y^{1-2s}\nabla V) = 0 & \text{in} \quad \mathbb{R}^{N+1}_+ \\ V(x, 0) = u^*(x) & \text{in} \quad \mathbb{R}^N. \end{cases} \tag{31}$$

It can be proved (see [7, Proposition 3.2]) that V has less energy than U^*, namely

$$I_1[V] + I_2[V] \leq I_1[U^*] + I_2[U^*], \tag{32}$$

where U solves (8). Therefore

$$[u^*]^2_{H^s(\mathbb{R}^N)} = I_1[V] + I_2[V] \leq I_1[U^*] + I_2[U^*] \leq I_1[U] + I_2[U] = [u]^2_{H^s(\mathbb{R}^N)}$$

and this proves Theorem 2.

Remark 13 By replacing s with $\frac{s}{2}$ and the boundary datum u with the characteristic function of some open and bounded set Ω in (9), the following characterization of the fractional perimeter can be used

$$P_s(\Omega) = c_{\frac{s}{2}} \inf_{\substack{U \in H^1_{\text{loc}}(\mathbb{R}^{N+1}_+) \\ U(\cdot,0)=\chi_\Omega}} \int_0^\infty y^{1-s} dy \int_{\mathbb{R}^N} \left(|\nabla_x U|^2 + |\partial_y U|^2\right) dx \tag{33}$$

and the fractional isoperimetric inequality through the classical Pólya-Szegö principle can also be achieved. The same representation formula holds for the first Dirichlet fractional eigenvalue just by using the first eigenfunction of $(-\Delta)^s$ as Dirichlet datum in the extension problem (8).

6 Stability of Some Geometric Inequalities in the Fractional Setting

Quantitative versions of functional inequalities aim to measure their *stability*, namely how far a given function or set is from attaining equality in the corresponding sharp inequality. In the fractional framework, such results are subtle due to the nonlocal nature of the underlying functionals. In this section, we review some noticeable results concerning sharp quantitative inequalities associated with nonlocal energies.

Concerning the fractional isoperimetric inequality, a quantitative enhancement involves the Fraenkel asymmetry

$$\mathcal{A}(\Omega) := \inf_{x\in\mathbb{R}^N} \frac{|\Omega\Delta B_r(x)|}{|\Omega|}, \quad \text{where } |B_r| = |\Omega|.$$

In general, given some geometric functional $\mathcal{F}$ and $m \in (0,\infty)$, once proved that the problem

$$\min_{|\Omega|=m} \mathcal{F}(\Omega)$$

admits a (unique) solution given by the ball, the goal is to prove that, under the measure constraint, the deficit $D(\Omega) := \mathcal{F}(\Omega) - \mathcal{F}(B)$ can be bounded from below by some positive function g of the Fraenkel asymmetry, namely

$$D(\Omega) \geq g(\mathcal{A}(\Omega)).$$

A very flexible technique in the local setting is the *Selection Principle* introduced in [18], which provides a way to extract a sequence $(E_h)_h$ of nearly optimal sets for a quantitative geometric inequality from arbitrary minimizing sequences. Then, using compactness and regularity results, one shows that $(E_h)_h$ converges (after suitable translations) to a ball, in the sense that

$$\mathcal{A}(E_h) \to 0 \quad \text{and} \quad \frac{D(E_h)}{g\left(\mathcal{A}(E_h)\right)} \to \inf_{\Omega} \frac{D(\Omega)}{g\left(\mathcal{A}(\Omega)\right)},$$

and satisfies uniform estimates of the second variations, which leads to the desired inequality. These tools have been adapted to the fractional perimeter in [31] obtaining the following quantitative form of (23).

Theorem 6 *For every $N \geq 2$ and $s_0 \in (0,1)$ there exists a positive constant $C_{N,s_0} > 0$ such that*

$$P_s(\Omega) \geq P_s(B) \left(\frac{|\Omega|}{|B|}\right)^{\frac{N-s}{N}} \left(1 + \frac{\mathcal{A}(\Omega)^2}{C_{N,s_0}}\right), \tag{34}$$

for every $s \in [s_0, 1]$ *and for every* $0 < |\Omega| < \infty$. *In particular, if* $|\Omega| = |B|$, *then*

$$D_s(\Omega) := P_s(\Omega) - P_s(B) \geq \frac{P_s(B)}{C_{N,s_0}} \mathcal{A}(\Omega)^2.$$

Remark 14 Some comments are in order. It can be proved via a counterexample that the power 2 in the right-hand side of (34) is sharp. Nevertheless, the Selection principle does not allow to find an explicit constant C_{N,s_0}, though it is conjectured that $C_{N,S_0} \approx \frac{C'_N}{s_0}$ as $s_0 \to 0^+$.

An alternative proof for quantitative isoperimetric inequalities exploits the extension technique stated in Theorem 1. In this framework, one can deduce a stability estimate basically using:

- the characterization of the $H^s(\mathbb{R}^N)$-seminorm (9)
- the sharp quantitative isoperimetric inequality by Fusco-Maggi-Pratelli [34]
- the so-called *transfer of asymmetry* which allows to estimate from below the asymmetry of the superlevel sets $E_{t,y} := \{x \in \mathbb{R}^N ; U_\Omega(x, y) > t\}$ of the extension U_Ω of the eigenfunction u_Ω for suitable levels $t > 0$ and suitable values of the extension variable $y \in (0, \infty)$ with a multiple of the asymmetry of Ω.

With this very flexible technique quantitative isoperimetric inqualities for the fractional perimeter in [35], for semilinear fractional Dirichlet eigenvalues [7] and for the fractional 2-capacity [19] have been obtained. Though the proof exploits the sharp quantitative isoperimetric inequality, the rearrangement of those inequalities on some extension levels does not lead to the sharp power of the Fraenkel asymmetry.

7 Some Generalizations

In this section we treat some generalizations when the relevant energy is not necessarily isotropic. Anisotropic energies reflect the influence of a non-Euclidean geometry on the variational structure of the problem. In this setting, symmetrization no longer leads to the euclidean ball as optimal shape but rather to Wulff shapes or other equivalent geometric structures, highlighting how the optimal shape inherits the underlying anisotropy.

7.1 The Anisotropic Case

The symmetrization of functionals of the form

$$\int_\Omega A(x, H(\nabla u))dx$$

for some norm $H : \mathbb{R}^N \to [0,\infty)$ and some Young function $A : \Omega \times [0,\infty) \to [0,\infty)$ has been addressed by several authors. We refer e.g. to the book [53].

A meaningful example is given by $A(x, H(\xi)) = H_K(\xi)$ for some norm H_K associated to a convex body $K \subset \mathbb{R}^N$. Moving to the fractional framework, anisotropic fractional seminorms have been introduced in [42], while the corresponding Pólya-Szegö principle with respect to the Steiner symmetrization defined in Sect. 2 for the relevant $W^{s,p}$ seminorm

$$[f]_{W^{s,p}_K(\mathbb{R}^N)} := \left(\int_{\mathbb{R}^N}\int_{\mathbb{R}^N} \frac{|u(x)-u(y)|^p}{\|x-y\|_K^{N+sp}}\,dx\,dy\right)^{1/p}$$

has been proved in [39] deducing an isoperimetric inequality for the relevant anisotropic fractional perimeter with minimizer given by the *Wulff shape* induced by K.

7.2 The Gaussian Case

Another significant example is given by $A(x, H(\xi)) = w(x)|\xi|^p$, for some $1 \le p < \infty$ and some positive weight w. In particular for $w = \gamma := \frac{e^{-\frac{|\cdot|^2}{2}}}{(2\pi)^{N/2}}$ we deal with the Gaussian rearrangement given by the *Ehrhard symmetrization* introduced in [27]. Through this symmetrization technique which rearranges a set in a halfspace keeping fixed its Gaussian volume, the same author proves the relevant Pólya-Szegö inequality [28, Théorème 3.1] and deduces the isoperimetric inequality [28, Proposition 2.1] and the Faber-Krahn inequality [28, Théorème 4.4]. See also [2, 15, 16] for quantitative estimates.

In the nonlocal counterpart the notion of H^s-fractional Gaussian seminorm has been introduced in [48] in the setting of abstract Wiener spaces through an extension technique *à la* Caffarelli-Silvestre proved in [51], and the authors prove an isoperimetric inequality for the *fractional Gaussian perimeter* $P_s^\gamma(E)$ using the characterization in formula (33) replacing $(\mathbb{R}^N, \mathcal{L}^N)$ with a Wiener space X equipped with the Gaussian measure $\gamma\mathcal{L}^N$. A stability result has also been obtained in [16] in finite dimension.

It is worth noticing that in the Gaussian framework the formulation with the extension technique is still equivalent with the one through Bochner subordination by considering the fractional powers of the (negative) *Ornstein-Uhlenbeck operator* $-\Delta_\gamma$, where $\Delta_\gamma := \Delta - \langle x, \nabla\rangle$ instead of the laplacian. See [51].

The usefulness of this approach is manifold; comparison results as in Theorem 5 for solutions to equations driven by fractional powers of $-\Delta_\gamma$ can be proved (see [29]). Moreover, it simplifies the study of $W^{s,p}_\gamma$ fractional seminorms *á la* Gagliardo or the analysis of the fractional Gaussian perimeter of a set in a bounded domain, see [13, 14].

Another, non-equivalent, definition of fractional perimeter in the Gauss space has been given in [23], where the authors prove that the only volume constrained isoperimetric sets are halfspaces passing through the origin, proving this way that for the given definition of Gaussian fractional perimeter the Ehrhard symmetrization may in general be not convenient.

8 Final Comments and Open Problems

Symmetrization methods are invaluable tools also in the fractional setting, allowing the derivation of sharp inequalities involving nonlocal geometric functionals. We conclude this chapter with some open problems

1. **Sharp quantitative inequalities** Many inequalities stated in Sect. 6 are conjectured to be not sharp. To improve them the use of the Selection Principle as done in [31], seems unavoidable.
2. **Pólya-Szegö inequalities for Dirichlet integrals of the Riesz fractional gradient** In [47, Section 4] the authors set as open problem the validity or not of the inequality

$$\int_{\mathbb{R}^N} |\nabla^s u^*|^p dx \leq \int_{\mathbb{R}^N} |\nabla^s u|^p dx. \tag{35}$$

being

$$\nabla^s u(x) := \mu_{N,s} \int_{\mathbb{R}^N} \frac{(u(x) - u(y))(x - y)}{|x - y|^{N+s+1}} dy$$

the *Riesz fractional gradient*, in the potential space $H^{s,p}(\mathbb{R}^N) := \{u \in L^p(\mathbb{R}^N) : \nabla^s u \in L^p(\mathbb{R}^N, \mathbb{R}^N)\}$. We notice that in general the equality

$$\|\nabla^s u\|_{L^p(\mathbb{R}^N,\mathbb{R}^N)} = C_{N,s,p}[u]_{W^{s,p}(\mathbb{R}^N)}$$

holds only for $p = 2$, and $H^{s,2}(\mathbb{R}^N) = H^s(\mathbb{R}^N)$. Then, for $p = 2$ inequality (35) immediately follows from Theorem 2, but for $p \neq 2$ neither the Schwarz symmetrization nor the Steiner symmetrization are not known to be useful. See e.g. [20] for further properties of $H^{s,p}(\mathbb{R}^N)$ and ∇^s.
3. **Fractional isoperimetric inequalities in the sub-Riemannian setting** One of the most celebrated open problems in sub-Riemannian geometry concerns the identification of the optimal isoperimetric profile. In the Heisenberg group $\mathbb{H}^N$, this question is known as *Pansu's conjecture*, which remains open despite several partial results. In contrast, for $\alpha \geq 0$ the metric measure space $(\mathbb{R}^{N+1}, d, \mathcal{L}^{N+1})$ with $ds^2 = |dx|^2 + \frac{dy^2}{(\alpha+1)|x|^{2\alpha}}$ is commonly known as *Grushin*

space and has an explicit isoperimetric profile which has been determined in [32, 46].
A natural question is to define a *fractional* analogue of the intrinsic distributional perimeter in Grushin spaces. A possible approach could rely on an extension procedure associated with the intrinsic laplacian given by the (negative) *Grushin operator* $-\mathcal{G}_\alpha$, where

$$\mathcal{G}_\alpha u := \Delta_x u + (\alpha+1)|x|^{2\alpha}\partial_{yy}u,$$

and on an adaptation of the Caffarelli–Silvestre-type extension to deal with the energy functional

$$\int_0^\infty z^{1-s}dz \int_{\mathbb{R}^{N+1}} |\nabla^\alpha_{x,y}U(x,y,z)|^2 + |\partial_z U(x,y,z)|^2 dxdy,$$

where

$$|\nabla^\alpha_{x,y}U(x,y,z)|^2 = |\nabla_x U(x,y,z)|^2 + (\alpha+1)^2|x|^{2\alpha}|\partial_y U(x,y,z)|^2.$$

In [3], the author observes that the Grushin metric is conformal to the Poincaré one through a change of variables. When endowed with the Poincaré metric the upper-half space $\mathbb{R}^{N+1}_+$ is known to be a model of hyperbolic geometry. In this setting the *hyperbolic symmetrization* rearranges the super-level sets of a function into metric balls keeping fixed the volume, so that it plays the role of the Schwarz symmetrization in the isotropic case and decreases the integral

$$\int_{\mathbb{R}^{N+1}} |\nabla^\alpha_{x,y}U(x,y,z)|^2 dxdy$$

for every $z > 0$, thus yielding that also the term

$$J_1[U] := \int_0^\infty z^{1-s}dz \int_{\mathbb{R}^{N+1}} |\nabla^\alpha_{x,y}U(x,y,z)|^2 dxdy$$

decreases applying the partial hyperbolic symmetrization with respect to the variable (x, y) to U. However, it is not clear how to deal with the term

$$J_2[U] := \int_0^\infty z^{1-s}dz \int_{\mathbb{R}^{N+1}} |\partial_z U(x,y,z)|^2 dxdy,$$

for which an intrinsic version of the Steiner symmetrization (as in [8]) would be required.

A rigorous implementation of this strategy could lead to a fractional isoperimetric inequality in the Grushin setting, yielding a different isoperimetric profile with respect to the one obtained for the classical distributional perimeter in [32, 46].

Acknowledgments The author kindly thanks Dr. Alberto Maione and Dr. Joaquim Duran for the opportunity of writing this book chapter. The author thanks also the Centre de Recerca Matematica for the very kind hospitality in March 2025.

Most of the results in this chapter were the subject of a PhD course held by the author in January 2025 at the Dipartimento di Matematica e applicazioni "Federico II" of the University of Naples for which the author thanks Prof. Carlo Nitsch.

The author also warmly thanks Prof. Diego Pallara and the anonymous referee for very useful suggestions in the revision of the chapter.

Funding Information The author is member of Gruppo Nazionale per l'Analisi Matematica, Probabilitá e le loro Applicazioni of the Istituto Nazionale di Alta Matematica (INdAM-GNAMPA) and he acknowledges the support of the MUR - PRIN 2022 project "Elliptic and parabolic problems, heat kernel estimates and spectral theory" (N. 20223L2NWK) and of the INdAM - GNAMPA 2025 Project "Metodi variazionali per problemi dipendenti da operatori frazionari isotropi e anisotropi" (Grant Code: CUP_E5324001950001). The author has no conflicts of interest to declare that are relevant to the content of this chapter.

References

1. Almgren, Jr., F.J., Lieb, E.H.: Symmetric decreasing rearrangement is sometimes continuous. J. Am. Math. Soc. **2**(4), 683–773 (1989)
2. Barchiesi, M., Brancolini, A., Julin, V.: Sharp dimension free quantitative estimates for the Gaussian isoperimetric inequality. Ann. Probab. **45**(2), 668–697 (2017)
3. Beckner, W.: On the Grushin operator and hyperbolic symmetry. Proc. Am. Math. Soc. **129**(4), 1233–1246 (2001)
4. Bourgain, J., Brezis, H., Mironescu, P.: Another look at Sobolev spaces. In: Optimal Control and Partial Differential Equations. IOS, Amsterdam (2001)
5. Brasco, L., Lindgren, E., Parini, E.: The fractional Cheeger problem. Interfaces Free Bound **16**, 419–458 (2014)
6. Brasco, L., Mosconi, S., Squassina, M.: Optimal decay of the extremals of the fractional Sobolev inequality. Calc. Var. PDE **55**(23), 32 pp. (2016)
7. Brasco, L., Cinti, E., Vita, S.: A quantitative stability estimate for the fractional Faber-Krahn inequality. J. Funct. Anal. **279**(3), Paper No. 108560, 49 pp. (2020)
8. Brock, F.: Weighted Dirichlet-type inequalities for Steiner symmetrization. Calc. Var. Partial Differ. Equ. **8**(1), 15–25 (1999)
9. Bucur, C., Valdinoci, E.: Nonlocal Diffusion and Applications. Lecture Notes of the Unione Matematica Italiana, vol. 20, Springer, Cham; Unione Matematica Italiana, Bologna (2016)
10. Caffarelli, L., Silvestre, L.: An extension problem related to the fractional Laplacian. Comm. Partial Differ. Equ. **32**(7–9), 1245–1260 (2007)
11. Caffarelli, L., Roquejoffre, J.-M., Savin, O.: Nonlocal minimal surfaces. Comm. Pure Appl. Math. **63**(9), 1111–1144 (2010)
12. Carbotti, A., Dipierro, S., Valdinoci, E.: Local Density of Solutions to Fractional Equations. De Gruyter Studies in Mathematics, 74. De Gruyter, Berlin (2019). ©2019
13. Carbotti, A., Cito, S., La Manna, D.A., Pallara, D.: Asymptotics of the s-fractional Gaussian perimeter as $s \to 0^+$. Fract. Calc. Appl. Anal. **25**(4), 1388–1403 (2022)
14. Carbotti, A., Cito, S., La Manna, D.A., Pallara, D.: Gamma-convergence of Gaussian fractional perimeter. Adv. Calc. Var. **16**(3), 571–595 (2023)
15. Carbotti, A., Cito, S., La Manna, D.A., Pallara, D.: Stability of the Gaussian Faber-Krahn inequality. Ann. Mat. Pura Appl. (4) **203**(5), 2185–2198 (2024)

16. Carbotti, A., Cito, S., La Manna, D.A., Pallara, D.: A quantitative dimension free isoperimetric inequality for the fractional Gaussian perimeter. Comm. Anal. Geom. **32**(2), 577–603 (2024)
17. Cesaroni, A., Novaga, M.: The isoperimetric problem for nonlocal perimeters. Discrete Contin. Dyn. Syst. Ser. S **11**(3), 425–440 (2018)
18. Cicalese, M., Leonardi, G.P.: A selection principle for the sharp quantitative isoperimetric inequality. Arch. Ration. Mech. Anal. **206**(2), 617–643 (2012)
19. Cinti, E., Ognibene, R., Ruffini, B.: A quantitative stability inequality for fractional capacities. Math. Eng. **4**(5), 1–28 (2022)
20. Comi, G.E., Stefani, G.: A distributional approach to fractional Sobolev spaces and fractional variation: existence of blow-up. J. Funct. Anal. **277**(10), 3373–3435 (2019)
21. Cotsiolis, A., Tavoularis, N.K.: Best constants for Sobolev inequalities for higher order fractional derivatives. J. Math. Anal. Appl. **295**(1), 225–236 (2004)
22. Dávila, J.: On an open question about functions of bounded variation. Calc. Var. Partial Differ. Equ. **15**(4), 519–527 (2002)
23. De Rosa, A., La Manna, D.A.: A nonlocal approximation of the Gaussian perimeter: gamma convergence and Isoperimetric properties. Commun. Pure Appl. Anal. **20**(5), 2101–2116 (2021)
24. del Teso, F., Gómez-Castro, D., Vázquez, J.L.: Three representations of the fractional p-laplacian: Semigroup, Extension and Balakrishnan Formulas. Fract. Calc. Appl. Anal. **24**, 966–1002 (2021)
25. Di Castro, A., Novaga, M., Ruffini, B., Valdinoci, E.: Nonlocal quantitative isoperimetric inequalities. Calc. Var. Partial Differ. Equ. **54**(3), 2421–2464 (2015)
26. Di Nezza, E., Palatucci, G., Valdinoci, E.: Hitchhiker's guide to the fractional Sobolev spaces. Bull. Sci. Math. **136**(5), 521–573 (2012)
27. Ehrhard, A.: Symétrisation dans l'espace de Gauss. Math. Scand. **53**(2), 281–301 (1983)
28. Ehrhard, A.: Inégalités isopérimétriques et intégrales de Dirichlet gaussiennes. Ann. Sci. Éc. Norm. Sup. (4) **17**(2), 317–332 (1984)
29. Feo, F., Stinga, P.R., Volzone, B.: The fractional nonlocal Ornstein-Uhlenbeck equation, Gaussian symmetrization and regularity. Discrete Contin. Dyn. Syst. **38**(7), 3269–3298 (2018)
30. Ferone, V., Piscitelli, G., Volzone, B.: Symmetrization results for general nonlocal linear elliptic and parabolic problems. J. Math. Pures Appl. (9) **190**, Paper No. 103597, 35 pp. (2024)
31. Figalli, A., Fusco, N., Maggi, F., Millot, V., Morini, M.: Isoperimetry and stability properties of balls with respect to nonlocal energies. Comm. Math. Phys. **336**(1), 441–507 (2015)
32. Franceschi, V., Monti, R.: Isoperimetric problem in H-type groups and Grushin spaces. Rev. Mat. Iberoam. **32**(4), 1227–1258 (2016)
33. Frank, R.L., Seiringer, R.: Non-linear ground state representations and sharp Hardy inequalities. J. Funct. Anal. **255**(12), 3407–3430 (2008)
34. Fusco, N., Maggi, F., Pratelli, A.: The sharp quantitative isoperimetric inequality. Ann. Math. **168**, 941–980 (2008)
35. Fusco, N., Millot, V., Morini, M.: A quantitative isoperimetric inequality for fractional perimeters. J. Funct. Anal. **261**(3), 697–715 (2011)
36. Garofalo, N.: Fractional thoughts. In: New Developments in the Analysis of Nonlocal Operators. Contemporary Mathematics, vol. 723, pp. 1–135. American Mathematical Society, Providence (2019)
37. Garofalo, N.: On the best constant in the nonlocal isoperimetric inequality of Almgren and Lieb. Atti Accad. Naz. Lincei Rend. Lincei Mat. Appl. **31**(2), 465–470 (2020)
38. Gilboa, G., Osher, S.J.: Nonlocal operators with applications to image processing. Multiscale Model. Simul. **7**(3), 1005–1028 (2008)
39. Kreuml, A.: The anisotropic fractional isoperimetric problem with respect to unconditional unit balls. Commun. Pure Appl. Anal. **20**(2), 783–799 (2021)
40. Lieb, E.H.: Existence and Uniqueness of the minimizing solution of Choquard's nonlinear equation. Stud. Appl. Math. **57**(2), 93–105 (1977)
41. Lombardini, L.: Fractional perimeters from a fractal perspective. Adv. Nonlinear Stud. **19**(1), 165–196 (2019)

42. Ludwig, M.: Anisotropic fractional perimeters. J. Differ. Geom. **96**(1), 77–93 (2014)
43. Martínez Carracedo, C., Sanz Alix, M.: The Theory of Fractional Powers of Operators. North-Holland Mathematics Studies, vol. 187, North-Holland Publishing Co., Amsterdam (2001)
44. Mazón, J.M., Rossi, J.D., Toledo, J.J.: Nonlocal Perimeter, Curvature and Minimal Surfaces for Measurable Sets. Frontiers in Mathematics. Birkhäuser/Springer, Cham (2019)
45. Maz' ya, V., Shaposhnikova, T.: On the Bourgain, Brezis, and Mironescu theorem concerning limiting embeddings of fractional Sobolev spaces. J. Funct. Anal. **195**(2), 230–238 (2002)
46. Monti, R., Morbidelli, D.: Isoperimetric inequality in the Grushin plane. J. Geom. Anal. **14**(2), 355–368 (2004)
47. Nguyen, H.-M., Squassina, M.: Some remarks on rearrangement for nonlocal functionals. Nonlinear Anal. **162**, 1–12 (2017)
48. Novaga, M., Pallara, D., Sire, Y.: A fractional isoperimetric problem in the Wiener space. J. Anal. Math. **134**(2), 787–800 (2018)
49. Pólya, G., Szegö, G.: Isoperimetric Inequalities in Mathematical Physics. Annals of Mathematical Studies. Princeton University Press, Princeton (1951)
50. Ponce, A.C.: A new approach to Sobolev spaces and connections to Γ-convergence. Calc. Var. Partial Differ. Equ. **19**(3), 229–255 (2004)
51. Stinga, P.R., Torrea, J.L.: Extension problem and Harnack's inequality for some fractional operators. Comm. Partial Differ. Equ. **35**(11), 2092–2122 (2010)
52. Valdinoci, E.: A fractional framework for perimeters and phase transitions. Milan J. Math. **81**(1), 1–23 (2013)
53. Van Schaftingen, J.: Anisotropic symmetrization. Ann. Inst. H. Poincaré C Anal. Non Linéaire **23**(4), 539–565 (2006)

Numerical Approximation of the Logarithmic Laplacian via Sinc-Basis

Patrick Dondl and Ludwig Striet

Abstract In recent works, the authors of this chapter have shown with co-authors how a basis consisting of dilated and shifted sinc-functions can be used to solve fractional partial differential equations. As a model problem, the fractional Dirichlet problem with homogeneous exterior value conditions was solved. In this work, we briefly recap the algorithms developed there and that—from a computational point of view—they can be used to solve nonlocal equations given through different operators as well. As an example, we numerically solve the Dirichlet problem for the logarithmic Laplacian $\log(-\Delta)$ which has the Fourier symbol $\log(|\omega|^2)$ and compute its Eigenvalues on disks with different radii in $\mathbb{R}^2$.

1 Introduction

Recently, nonlocal equations have been a widely studied topic in pure and applied mathematics. A prototypical example for such an equation is: find u such that

$$\begin{aligned} (-\Delta)^s u &= f \text{ in } \Omega \\ u &= 0 \text{ in } \mathbb{R}^d \setminus \Omega. \end{aligned} \tag{1}$$

Above, f is a right-hand side defined on an open, bounded domain Ω. For $s \in (0, 1)$, the operator $(-\Delta)^s$ is the *integral fractional Laplacian* which can be defined as

$$(-\Delta)^s u(x) := -C(d, s)\text{P.V.} \int_{\mathbb{R}^d} \frac{u(x) - u(y)}{|x - y|^{d+2s}} \mathrm{d}y.$$

P. Dondl (✉) · L. Striet
University of Freiburg, Department for Applied Mathematics, Freiburg, Germany
e-mail: patrick.dondl@mathematik.uni-freiburg.de; ludwig.striet@mathematik.uni-freiburg.de

J. Duran i Lamiel, A. Maione (eds.), *New Frontiers in Homogenization and Fractional Calculus*, Trends in Mathematics,
https://doi.org/10.1007/978-3-032-20098-3_4

This definition exposes major challenges in the development of numerical methods for problem of the form of (1): to evaluate the operator $(-\Delta)^s u(x)$ at a single point $x \in \mathbb{R}^d$, a singular integral over the whole $\mathbb{R}^d$ has to be computed. Furthermore, the definition exposes why conditions on u have to be imposed on $\mathbb{R}^d \setminus \Omega$ instead of only on the boundary of Ω, thus making them *exterior value conditions.*

Driven by the number of applications of such models, many numerical methods to solve equations of that kind have been proposed in literature recently.

This includes, among others, finite difference methods such as presented in [15, 17, 18]. In [24], the authors derive the fractional Laplacian as the limit of the long-jump random walk on a grid. This could also be used to derive the weights $w_k = |k|^{-(d+2s)}$ for a finite difference scheme. Collocation methods have been presented in [9, 22, 25]. Another large class of methods are Galerkin methods. Regularity theory and implementation details for a $\mathbb{P}^1$ finite element method for (1) have been given in [1]. More aspects of such methods, including efficient implementations, have been provided in [2, 3]. In [7], the authors present a non-conforming finite element method that amounts in solving a set of elliptic problems on accordingly truncated domains. In [12], the authors show that exponential convergence can be achieved when using *hp* finite elements in polygonal domains. In [13], the authors show that Duffy transforms can be used to overcome the difficulties that arise when computing the stiffness matrix for (1) in $d = 3$ dimensions.

In this chapter, we focus on an approach employed and analyzed in [5, 6], where it has been shown that a basis consisting of dilated and shifted sinc-functions can be used to approximate nonlocal problems with Dirichlet exterior value conditions when the operator is given as a Fourier symbol. The method presented in these articles can be seen either a collocation method or as a Galerkin method. The method relies on the fact that the operator $(-\Delta)^s$—acting on functions given on $\mathbb{R}^d$—can be equivalently defined as a Fourier multiplier. More precisely, the identity

$$\mathcal{F}\{(-\Delta)^s u\} = |\omega|^{2s} (\mathcal{F}u)(\omega) \tag{2}$$

holds. For a proof, see e.g. [24] or [19] where the equivalence of ten different definitions is shown.

In theory, the definition in (2) could be used directly to construct a numerical method via the discrete Fourier transformation as seen, e.g., in [4]. However, this introduces an implicit periodization of the function which may have undesired effects if homogeneous exterior value conditions as in (1) are desired.

The aforementioned approach presented in [5, 6] differs from that as sinc-functions are used as a basis for the numerical approximations. Those functions combine (i) a reasonable decay in the physical space that can appropriately model the exterior value conditions and (ii) a simple Fourier transformation which is essentially the indicator function of a box. While the focus of [5] is the efficient implementation of the method along with numerical experiments, the focus of [6] are numerical analysis and a convergence proof for the discretization of [6] under mild assumptions on Ω and f. We summarize the details that are important for this article in Sect. 3.

In this note, we show that—from a computational point of view—the developed sinc-function based techniques can be used to approximate a much broader range of problems. More specifically, we can compute an operator L as long it is defined through a Fourier symbol $m(\omega)$ via

$$\mathcal{F}\{Lu\} = m(\omega)\,(\mathcal{F}u)\,(\omega) \tag{3}$$

and solve associated (nonlocal) partial differential equations in the sense of collocation methods.

As an example, we numerically study some of the properties of the *logarithmic Laplacian* $\log(-\Delta)$ which can be defined via the identity

$$\mathcal{F}\{\log(-\Delta)u\} = \log(|\omega|^2)(\mathcal{F}u)(\omega).$$

The logarithmic Laplacian formally arises as the derivative $\partial_s\big|_{s=0}(-\Delta)^s$ of the fractional Laplacian at $s = 0$ [11]. This operator has been subject to investigations in articles recently. In [11], the authors study the Dirichlet problem for the logarithmic Laplacian and the spectral properties of the logarithmic Laplacian are subject of, among others, [20] and bounds for the eigenvalues are given in [10].

Recently, the numerical approximation of the logarithmic Laplacian on an interval has been studied in [16]. In this work, the authors show the implementation and analysis of a finite element method and they establish error estimates in appropriately defined weighted function spaces. The analysis is substantiated by numerical experiments. Furthermore, they show that the eigenvalues of the discretized stiffness matrix converge to the eigenvalues of the logarithmic Laplacian as the spatial discretization parameter h approaches 0.

In this work, we present an approach suitable for dimensions higher than one, obtaining results that are in good alignment with theoretical predictions. A rigorous numerical analysis of our method is an open problem and subject of ongoing work.

2 Mathematical Preliminaries

We introduce some of the basic facts regarding the Fourier transform that are required in the following sections. For details, we refer to the many textbooks on the topic such as, e.g., [14].

In this article, we use the conventions

$$\mathcal{F}u(\omega) = (2\pi)^{-d}\int_{\mathbb{R}^d} u(x)\mathrm{e}^{-i\omega x}\,\mathrm{d}x, \qquad \mathcal{F}^{-1}\hat{u}(x) = \int_{\mathbb{R}^d} \hat{u}(\omega)\mathrm{e}^{\mathrm{i}\omega x}\,\mathrm{d}\omega.$$

for the Fourier transform on $L^2(\mathbb{R}^d)$ and its inverse. The definition is considered to be extended to the space of tempered distributions accordingly where needed. An important tool in Fourier analysis that we make use of is the Fourier scaling theorem which states that the identity

$$\mathcal{F}(u(x/h))(\omega) = |h|^d \mathcal{F}u(h\omega)$$

holds for $h \in \mathbb{R}$.

Initially, our algorithms are implemented so that they work on a domain $\Omega \subset (0, 1)^d$. By translational invariance, we can easily see that no change is required to consider $\Omega \subset (-1/2, 1/2)^d$. Arbitrarily large domains can then be treated by scaling the Fourier symbols, as justified by the following lemma.

Lemma 1 *Let L a linear operator with symbol $m(\omega)$ and $L_{r/R}$ the operator with symbol $m(r/R\omega)$. Let u solve*

$$\begin{aligned} L_{r/R}u &= \lambda u \textit{ in } B_r \\ u &= 0 \textit{ in } \mathbb{R}^d \setminus B_r \end{aligned}$$

and $v(x) = u(r/Rx)$. Then v solves

$$Lv = \lambda v \textit{ in } B_R \tag{a}$$

$$v = 0 \textit{ in } \mathbb{R}^d \setminus B_R \tag{b}$$

Proof The second part (b) is obvious. The first part (a) is shown by a simple computation: let $x \in B_R$, then $r/Rx \in B_r$ and

$$\begin{aligned} \lambda v(x) &= \lambda u(r/Rx) = L_{r/R}u(r/Rx) \\ &= \mathcal{F}^{-1}\{m(r/R\omega)\hat{u}(\omega)\}(r/Rx) \\ &= \int_{\mathbb{R}^d} m(r/R\omega)\hat{u}(\omega)\mathrm{e}^{\mathrm{i}r/Rx\cdot\omega}\mathrm{d}\omega \\ &= \int_{\mathbb{R}^d} m(\omega)\hat{u}(R/r\omega)\mathrm{e}^{\mathrm{i}x\omega}(r/R)^{-d}\mathrm{d}\omega \\ &= \int_{\mathbb{R}^d} m(\omega)\hat{v}(\omega)\mathrm{e}^{\mathrm{i}x\omega}\mathrm{d}\omega \\ &= Lv(x) \end{aligned}$$

where we used that $\hat{v}(\omega) = (R/r)^d\hat{u}(R/r\omega)$ as a consequence of the Fourier scaling theorem. □

Similarly, we obtain the following lemma.

Lemma 2 *Let L a linear operator with symbol $m(\omega)$ and $L_{r/R}$ the operator with symbol $m(r/R\omega)$. Let u solve*

$$L_{r/R}u = f \; in \; B_r$$
$$u = 0 \; in \; \mathbb{R}^d \setminus B_r$$

and $v(x) = u(r/Rx)$. *Then* v *solves*

$$Lv = f(r/Rx) \; in \; B_R$$
$$v = 0 \; in \; \mathbb{R}^d \setminus B_R$$

Proof Same as the proof of Lemma 1, starting with $f(r/Rx) = L_{r/R}u(r/Rx)$ for $x \in B_R$. □

3 The Sinc-Method for Nonlocal Operators

In this section, we briefly recap the details needed for the implementation of the method presented in [5] to solve nonlocal equations.

The sinc-function is defined as

$$\operatorname{sinc}(x) := \frac{\sin(\pi x)}{\pi x} = \int_{[-\pi,\pi]} \frac{1}{2\pi} e^{i\omega\cdot x} d\omega = \mathcal{F}^{-1}\left\{\frac{1}{2\pi}\chi_{[-\pi,\pi]}\right\} \tag{4}$$

where

$$\chi_{[-\pi,\pi]}(\omega) = \begin{cases} 1 & \text{if } \omega \in [-\pi,\pi] \\ 0 & \text{otherwise} \end{cases}.$$

We notice from (4) that the sinc function is obtained as the inverse Fourier transform of the indicator function of the interval $[-\pi, \pi]$ in $\mathbb{R}$.

To approximate problems in the form of (1) and, more general, operators of the form of (3), we use dilated and shifted tensor products of the sinc-function defined in (4). Namely, we define the basis function $\varphi(x) = \operatorname{sinc}(x)$ and for a positive integer N and a multiindex $k = (k_0, \dots, k_d) \in \{0, \dots, N-1\}^d$, we define the function

$$\varphi_k^N(x) := \prod_{i=1}^{d} \varphi(Nx - k_i).$$

For a multiindex $k \in \mathbb{Z}^d$, we define the grid points $x_k = k/N$. Note that the basis functions fulfill the property

$$\varphi_k^N(x_j) = \delta_{k,j} = \begin{cases} 1 & \text{if } k = j \\ 0 & \text{otherwise} \end{cases}.$$

We define the discrete, finite dimensional function space

$$\mathbb{V}_h(\Omega) = \left\{ v_h(x) = \sum_{k \in \mathbb{Z}^d} v_k \varphi_k^N(x) \mid v_k \in \mathbb{R},\, v_k = 0 \text{ if } x_k \notin \Omega \right\}$$

and solve the discrete equation: find $u_h \in \mathbb{V}_h(\Omega)$ that fulfills

$$Lu_h(x_k) = f(x_k). \tag{5}$$

To solve (5), we have to discretize the operator L. The details are provided in great detail in [5, 23]. Briefly, the idea is the following. Evaluating $Lv_h(x_\kappa)$ for $v_h \in V_h(\Omega)$ and a grid point x_κ results in computing

$$\begin{aligned}
(-\Delta)^s v_h(x_\kappa) &= L\Big(\sum_{k \in \Omega_h} v_k \varphi_k^N(x_\kappa) \Big) \\
&= \sum_{k \in \Omega_h} v_k \underbrace{\left(L\varphi_k^N \right)(x_\kappa)}_{=:\Phi^N(\kappa - k)} \\
&= \sum_{k \in I_N^d} v_k \Phi^N(\kappa - k).
\end{aligned}$$

with $I_N = \{0, \ldots, N-1\}$. This is a discrete convolution and can be evaluated efficiently using the discrete fourier transform once $\Phi^N(\kappa - k)$ is known for all $\kappa - k$. The authors show in the aforementioned works that while evaluating $\Phi_k^N = L\varphi_k^N$ directly is hard, computing its discrete Fourier transform can be done as follows: Using the fact that

$$\Phi^N(\kappa - k) = L\varphi_k^N(x_\kappa) = L\varphi_k^N(x_\kappa - x_k) = \mathcal{F}^{-1}\left(m(\omega)\hat{\varphi}^N(\omega) \right)(x_\kappa - x_k)$$

and the definition of the discrete Fourier transform of size $(2N)^d$, we know for the k-th coefficient of the discrete Fourier transform of Φ that it can be obtained through the equation

$$\widehat{\Phi}_k^N = (2\pi)^{-d} \int_{[-N,N]^d} m(\pi\omega) Y_d\left(\frac{\pi}{N}(\omega - k) \right) \mathrm{d}\omega$$

where, for $\omega = (\omega_1, \ldots, \omega_d)^T$,

$$Y_d(\omega) = \prod_{i=1}^d Y(\omega_i), \quad Y(x) := \sum_{j=-N}^{N-1} \mathrm{e}^{\mathrm{i}jx} = \begin{cases} \frac{\mathrm{e}^{-\mathrm{i}Nx}(\mathrm{e}^{2\mathrm{i}Nx} - 1)}{\mathrm{e}^{\mathrm{i}x} - 1} & \text{if } \mathrm{e}^{\mathrm{i}x} - 1 \neq 0 \\ 2N & \text{otherwise} \end{cases}.$$

This is still an osciallating integral, but it can be computed approximately as follows. First, we split the integral up into $(2N)^d$-many integrals over cubes with side length 1 and obtain

$$\widehat{\Phi}_k^N = \sum_{j \in i_{2N}'^d} \int_{Q_j} m(\pi\omega) Y_d \left(\frac{\pi}{N}(\omega - k)\right) \mathrm{d}\omega \tag{6}$$

where $I'_{2N} := \{-N, \dots N-1\}$ and $Q_j := [j_1, j_1+1] \times \cdots \times [j_d, j_d+1]$. To evaluate this, we choose a quadrature rule $(x_i, \alpha_i)_{i=1,\dots,N_Q}$ is a quadrature rule on $[0,1]^d$, apply it on each of the cubes and compute the values of $\widehat{\Phi}_k^N$ via the formula

$$\widehat{\Phi}_k^N \approx (2N)^{-d} \sum_{i=1}^{N_Q} \alpha_i \sum_{j \in I_{2N}'^d} m(\pi(j + x_i)) Y_d \left(-\frac{\pi}{N}(k - j) + \frac{\pi}{N} x_i\right) \tag{7}$$

with $I'_{2N} = \{-N, \dots, N-1\}$. Here, each of the inner sums has the structure of a discrete convolution again, which can be used to implement (7) efficiently. This is still computationally demanding, but has to be done only once for each N and each operator.

To implement the above approach, one has to choose a quadrature rule. In principle, there is no particular restriction, except that the same rule has to be used on each cube. Tensor Gauß-Legendre rules have been used for the relevant implementations in [5] and it has been seen in numerical experiments that this is a legitimate choice for the fractional Laplacian, even though the symbol exhibits reduced regularity near the origin, making evaluation of (7) more difficult for multi-indices j where one or more components are 0 or -1.

In the case of the logarithmic Laplacian, the expression in (7) has an even more pronounced singularity near the origin which thus should be treated properly. This can be achieved for example by means of a Duffy transform, which in our case consists of a singular domain transformation to cancel a singularity of an integrand at a corner of the domain.

We illustrate the procedure by computing the integral

$$\int_{Q_0} m(\pi\omega) Y_d \left(\frac{\pi}{N}\omega\right) \mathrm{d}\omega = \int_0^1 \int_0^1 m(\pi\omega_1, \pi\omega_2)\, y_0(\omega_1, \omega_2) \mathrm{d}\omega_2 \mathrm{d}\omega_1$$

where we abbreviate

$$y_0(\omega_1, \omega_2) = Y_d \left(\frac{\pi}{N}\omega\right).$$

This is precisely the integral over the 0th cube $Q_0 = (0,1)^2$ for $k = (0,0)^T$ in (6). It is clear that the integrand has the aforementioned singularity at $(\omega_1, \omega_2) = (0,0)$ for $m(\omega_1, \omega_2) = \log(\omega_1^2 + \omega_2^2)$. The idea of the Duffy transform is to split the integral over the cube into two integrals over triangles via

$$
\begin{aligned}
&\int_0^1 \int_0^1 m(\pi\omega_1, \pi\omega_2)\, y_0(\omega_1, \omega_2) \mathrm{d}\omega_2 \mathrm{d}\omega_1 \\
&= \int_0^1 \int_0^{\omega_1} m(\pi\omega_1, \pi\omega_2)\, y_0(\omega_1, \omega_2) \mathrm{d}\omega_2 \mathrm{d}\omega_1 \\
&\quad + \int_0^1 \int_0^{\omega_2} m(\pi\omega_1, \pi\omega_2)\, y_0(\omega_1, \omega_2) \mathrm{d}\omega_1 \mathrm{d}\omega_2. \qquad (8)
\end{aligned}
$$

so that the singularity is in the corner of one of them. Then, the singularity in the first integral is removed by transforming it to an integral over a cube again as

$$
\begin{aligned}
&\int_0^1 \int_0^{\omega_1} m(\pi\omega_1, \pi\omega_2)\, y_0(\omega_1, \omega_2) \mathrm{d}\omega_2 \mathrm{d}\omega_1 \\
&= \int_0^1 \int_0^1 m(\pi\omega_1, \pi\omega_1\eta)\, y_0(\omega_1, \omega_1\eta)\omega_1 \mathrm{d}\omega_1 \mathrm{d}\eta.
\end{aligned}
$$

The second integral is transformed the same way as

$$
\begin{aligned}
&\int_0^1 \int_0^{\omega_2} m(\pi\omega_1, \pi\omega_2)\, y_0(\omega_1, \omega) \mathrm{d}\omega_1 \mathrm{d}\omega_2 \\
&= \int_0^1 \int_0^1 m(\pi\omega_2\eta, \pi\omega_2)\, y_0(\omega_2\eta, \omega)\omega_2 \mathrm{d}\eta \mathrm{d}\omega_2
\end{aligned}
$$

and we obtain

$$
\begin{aligned}
&\int_0^1 \int_0^1 m(\pi\omega_1, \pi\omega_2)\, y_0(\omega_1, \omega) \mathrm{d}\omega_2 \mathrm{d}\omega_1 \\
&= \int_0^1 \int_0^1 m(\pi\omega_1, \pi\omega_1\eta)\, y_0(\omega_1, \omega_1\eta)\omega_1 \mathrm{d}\omega_1 \mathrm{d}\eta \\
&\quad + \int_0^1 \int_0^1 m(\pi\omega_2\eta, \pi\omega_2)\, y_0(\omega_2\eta, \omega)\omega_2 \mathrm{d}\eta \mathrm{d}\omega_2 \\
&= \int_0^1 \int_0^1 \big(m(\pi\omega, \pi\eta\omega) y(\omega, \eta\omega) + m(\pi\eta\omega, \pi\omega) y(\eta\omega, \omega)\big)\omega \mathrm{d}\omega \mathrm{d}\eta. \qquad (9)
\end{aligned}
$$

by plugging both into (8), renaming $\omega_1 = \omega$ in the first integral, $\omega_2 = \omega$ and changing the order of integration in the second integral and summarizing both into one integral over $(0, 1)^2$.

As $Y(\cdot)$ is smooth and bounded, the singularity results only from $m(\omega) = \log(|\omega|^2)$. The transform as described in (9) transforms this singularity into an expression of the form

$$\omega m(\pi\omega, \pi\omega\eta) = \omega \log\left((\pi\omega)^2 + (\pi\omega\eta)^2\right) = \omega \log\left(\pi^2(1+\eta^2)\omega^2\right)$$

which remains bounded and can be integrated using standard quadrature formula.

However, as this affects only the cubes adjacent to the origin, we only want to apply an appropriate rule on these cubes, but we have to apply the same rule to all cubes in order to compute the coefficients via (7) which is necessary for an efficient algorithm.

We solve this issue by first computing $\widehat{\Phi}_k^N$ for all k via (7). Then, we subtract the summands that belong to cubes near the origin from each of the coefficients and compute those again using a quadrature rule on the integral transformed to the form of (9).

4 Numerical Experiments

In this section, we show the results of some new numerical experiments. This includes numerical error rates for the Dirichlet problem with integral fractional Laplacian in $d = 4$ spatial dimensions in Sect. 4.1, and experiments regarding the eigenvalues of the logarithmic Laplacian on the ball in Sect. 4.2 and the Dirichlet problem for the logarithmic Laplacian on the ball in Sect. 4.3.

4.1 The Dirichlet Problem for the Fractional Laplacian on the Ball

A standard problem to test numerical methods for the integral fractional Laplacian is to find u that fulfills

$$\begin{aligned} (-\Delta)^s u &= 1 \text{ in } B_1 \\ u &= 0 \text{ in } \mathbb{R}^d \setminus B_1 \end{aligned} \tag{10}$$

which has the analytic solution

$$u = C_u(d,s) \max\left(0, (1-|x|^2)\right)^s, \quad C_u(d,s) = \frac{\Gamma(d/2)}{2^{2s}\Gamma(d/2+s)\Gamma(1+s)},$$

see e.g. [8] also for the exact value of $C_u(d,s)$. Error decay rates for the sinc-method for this problem in the energy norm and in the L^2 norm have been computed numerically in [5] in $d = 2$ and $d = 3$ spatial dimensions and proven later in [6] for arbitrary spatial dimensions. Although the proofs are only for the error decay rate in

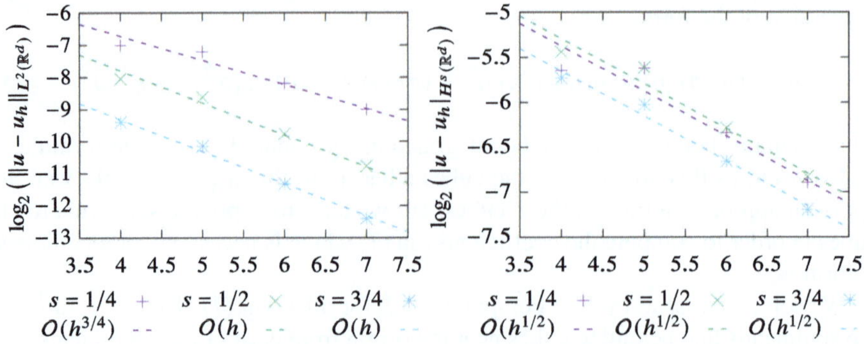

Fig. 1 The numerical errors that we compute for the problem in (10) in $d = 4$ spatial dimensions in the H^s seminorm (left) and in the L^2 norm (right). The dashed lines indicate that we obtain the same error decay rates as in $d = 2$ and $d = 3$

the energy norm, the numerically computed decay rates in L^2 match well with what has been proven for finite element methods and are $\sim h^{\min(1/2+s,1)}$.

In principle, the method can be implemented in arbitrary spatial dimensions, although the "curse of dimensionality" is a problem in higher dimensions. Still, we show numerical error decay rates in the L^2-norm in $d = 4$ with h up to 2^{-7} which corresponds to approximately $54.3 \cdot 10^6$ unknowns.

We show the errors that we compute in Fig. 1. The results suggest that we obtain the same error decay rates as in $d = 2$ and $d = 3$ spatial dimensions.

4.2 The Eigenvalues of the Logarithmic Laplacian on the Ball

As has been noted in [20], the eigenvalues of the logarithmic Laplacian with Dirichlet boundary conditions on the unit ball B_R depend on the radius R. Furthermore, for specific radii R_ℓ, zero is an eigenvalue of the logarithmic Laplacian. The eigenvalues of the logarithmic Laplacian on the ball B_R are then

$$\lambda^{(\ell)} = 2\log(R/R_\ell), \tag{11}$$

see [20, Lemma 2.5], [16].

In a first numerical experiment, we aim to approximately compute the smallest eigenvalues of the approximation of the Dirichlet logarithmic Laplacian on B_R for different values of $R > 0$. Once known, these can be used to simplify implementations, we seek the eigenvalues of the operator with symbol $\log\left(|r/R\omega|^2\right)$ on $B_r \subset (0, 1)^2$ to obtain the eigenvalues of the (approximated) logarithmic Laplacian with symbol $2\log(|\omega|)$ on $B_R \subset \mathbb{R}^d$ as justified by Lemma 1.

To compute the eigenvalues $\lambda_h^{(\ell)}$, we setup the operator with $N = 2^{10}$ grid points in each spatial direction and a 7×7-point tensor Gauß-Legendre rule. We use the Spectra Library [21] to compute the eigenvalues of Φ^N then.

We remark that we are currently unable to provide a proof that the computed eigenvalues indeed converge to the exact eigenvalues of the logarithmic Laplacian. However, it is worth to note that the pairs of eigenvalues $\lambda_h^{(\ell)}$ and eigenvectors $\mathbf{v}_h^{(\ell)}$ that fulfill $(\Phi^N \mathbf{v}^{(\ell)})_k = (\lambda_h \mathbf{v}^{(\ell)})_k$ for $k \in \Omega_N$ do represent eigenvalues and eigenfunctions of the logarithmic Laplacian in the following sense: Let $v_h^{(\ell)}$ the sinc-function associated with the the coefficient vectors $\mathbf{v}^{(\ell)}$, then

$$\log(-\Delta) v_h(x_k) = (\Phi^N \mathbf{v}^{(\ell)})_k = (\lambda_h^{(\ell)} \mathbf{v}^{(\ell)})_k = \lambda_h^{(\ell)} v_h^{(\ell)}(x_k) \quad \text{for } x_k \in B_R.$$

We show the eigenvalues that we compute for different values of R in Fig. 2, where we also see that the eigenvalues scale logarithmically as expected.

In a next numerical experiment, we aim to seek the exact values of R when 0 is an eigenvalue of the logarithmic Laplacian on B_R with Dirichlet exterior value conditions. These values could be used to estimate the Eigenvalues on balls with arbitrary radii via (11).

To find the radius for which the ℓth eigenvalue is 0, we take (i) the biggest R for which $\lambda^{(\ell)}$ is greater than 0 and (ii) the smallest R for which $\lambda^{(\ell)}$ is smaller than 0 from Fig. 2 and iteratively refine them until we find R_ℓ so that $\lambda^{(\ell)} = 0$. We show the values of R_ℓ in Table 1. Furthermore, we show some of the eigenfunctions that belong to the respective 0-eigenvalue in Fig. 3.

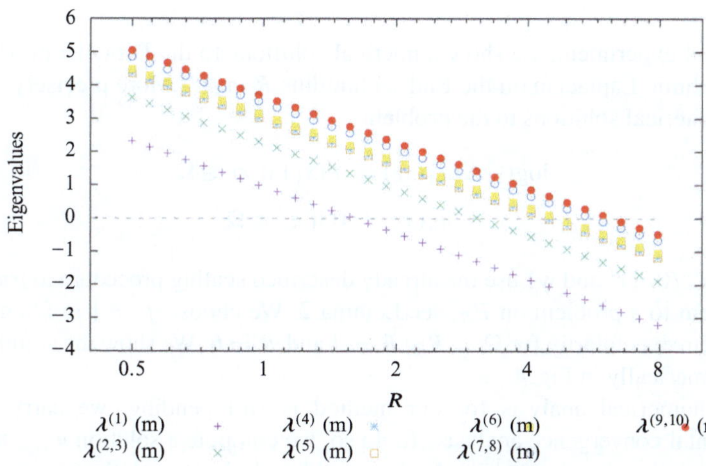

Fig. 2 The first Dirichlet eigenvalues numerically computed on $B_R \subset \mathbb{R}^2$ using the sinc method. We notice that the eigenvalues scale logarithmically as expected

Table 1 The values of R_ℓ for which the ℓth eigenvalue is 0, rounded to 4 decimal places

ℓ	1	2	3	4	5	6	7	8	9	10
R_ℓ	1.6015	3.0910	3.0910	4.4221	4.4251	4.7248	5.6846	5.6846	6.2476	6.2476

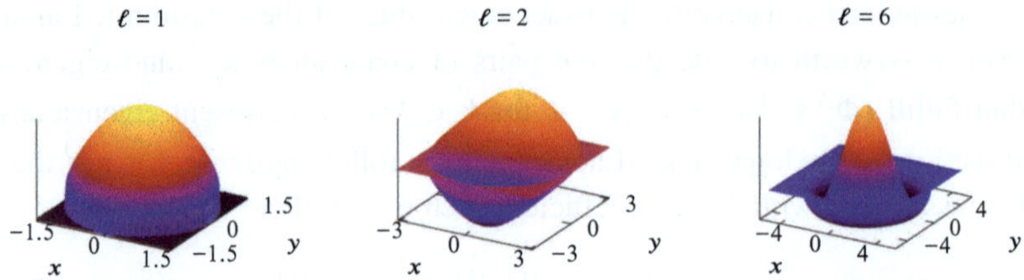

Fig. 3 The eigenfunctions that belong to the eigenvalue $\lambda_\ell = 0$ for the radii R_ℓ in Table 1

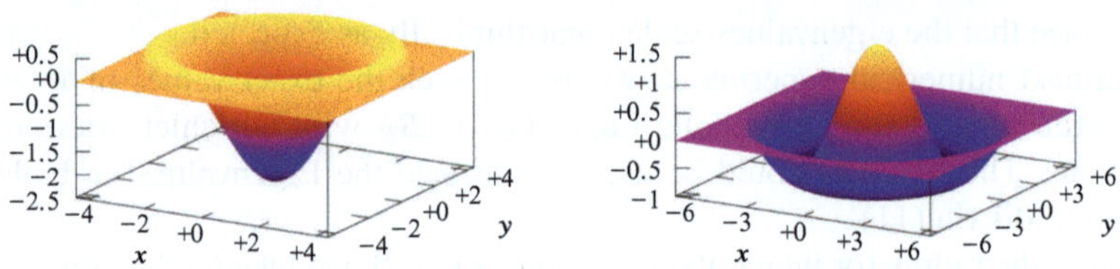

Fig. 4 The solutions we obtain to the Dirichlet problem for the logarithmic Laplacian on balls with radius $R = 4$ (left) and $R = 6$ (right). As expected, the solutions oscillate for larger values of R

4.3 The Dirichlet Problem for the Logarithmic Laplacian

In a second experiment, we show numerical solutions to the Dirichlet problem for the logarithmic Laplacian on the Ball with radius R in $\mathbb{R}^d$. More precisely, we aim to find numerical solutions to the problem

$$\log(-\Delta)u_h(x_k) = f(x_k) \text{ if } x_k \in \Omega$$
$$u_h(x_k) = 0 \text{ if } x_k \notin \Omega$$

where $\Omega \subset (0, 1)^d$ and we use the already described scaling procedure to transform the problem to a problem on B_R, see Lemma 2. We choose $f = 1$ in Ω and solve the problem exemplarily for $\Omega = B_R$, $R = 4$ and $R = 6$. We show the solutions we obtain numerically in Fig. 4.

As a numerical analysis for our method is still pending, we carry out an experimental convergence analysis. To do so, we compute a solution $u_{h_{\min}}$ at a fine spatial resolution $h_{\min} = 2^{-12}$ and compare this solution to solutions u_h computed at coarser solutions. We approximate the L^2-error as

$$\left\|u_h - u_{h_{\min}}\right\|_{L^2(\mathbb{R}^d)} \approx \left(h^2 \sum_{k \in \mathbb{Z}^d} (u_h(x_k) - u_{h_{\min}}(x_k))^2\right)^{1/2}.$$

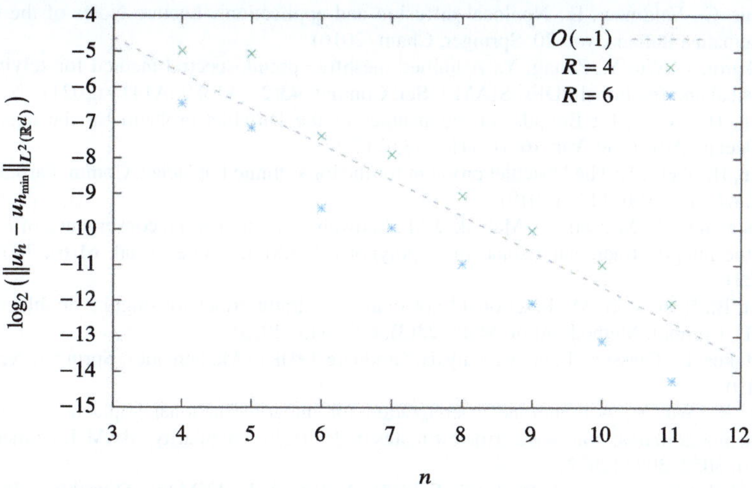

Fig. 5 The approximately computed $\|\cdot\|_{L^2(\mathbb{R}^d)}$ errors where u_h is the numerical solution obtained with spatial resolution $h = 2^{-n}$ and $u_{h_{\min}}$ is the numerical solution obtained at the minimal spatial resolution $h_{\min} = 2^{-12}$. For this experiment, the error rate is approximately -1 as indicated by the dashed grey line

The error decays with a rate of approximately h^{-1} in this experimental analysis as shown in Fig. 5.

Funding Information Ludwig Striet gratefully acknowledges a doctoral scholarship from the Friedrich-Ebert-Stiftung.

References

1. Acosta, G., Borthagaray, J.P.: A fractional Laplace equation: regularity of solutions and finite element approximations. SIAM J. Numer. Anal. **55**(2), 472–495 (2017)
2. Ainsworth, M., Glusa, C.: Aspects of an adaptive finite element method for the fractional Laplacian: a priori and a posteriori error estimates, efficient implementation and multigrid solver. Comput. Methods Appl. Mech. Engrg. **327**, 4–35 (2017)
3. Ainsworth, M., Glusa, C.: Towards an efficient finite element method for the integral fractional Laplacian on polygonal domains. In: Dick, J., Kuo, F.Y., Woźniakowski, H. (eds.) Contemporary Computational Mathematics, pp. 17–57. Springer, Cham (2018)
4. Antil, H., Bartels, S.: Spectral approximation of fractional PDEs in image processing and phase field modeling. Comput. Methods Appl. Math. **17**, 661–678 (2017)
5. Antil, H., Dondl, P., Striet, L.: Approximation of integral fractional Laplacian and fractional PDEs via sinc-basis. SIAM J. Sci. Comput. **43**(4), A2897–A2922 (2021)
6. Antil, H., Dondl, P.W., Striet, L.: Analysis of a sinc-Galerkin method for the fractional Laplacian. SIAM J. Numer. Anal. **61**(6), 2967–2993 (2023)
7. Bonito, A., Lei, W., Pasciak, J.E.: Numerical approximation of the integral fractional Laplacian. Numer. Math. **142**(2), 235–278 (2019)

8. Bucur, C., Valdinoci, E.: Nonlocal diffusion and applications. Lecture Notes of the Unione Matematica Italiana, vol. 20. Springer, Cham (2016)
9. Burkardt, J., Wu, Y., Zhang, Y.: A unified meshfree pseudospectral method for solving both classical and fractional PDEs. SIAM J. Sci. Comput. **43**(2), A1389–A1411 (2021)
10. Chen, H., Véron, L.: Bounds for eigenvalues of the Dirichlet problem for the logarithmic Laplacian. Adv. Calc. Var. **16**(3), 541–558 (2022)
11. Chen, H., Weth, T.: The Dirichlet problem for the logarithmic Laplacian. Comm. Partial Differ. Equ. **44**(11), 1100–1139 (2019)
12. Faustmann, M., Marcati, C., Melenk, J.M., Schwab, C.: Exponential convergence of hp-FEM for the integral fractional Laplacian in polygons. SIAM J. Numer. Anal. **61**(6), 2601–2622 (2023)
13. Feist, B., Bebendorf, M.: Fractional Laplacian – quadrature rules for singular double integrals in 3D. Comput. Methods Appl. Math. **23**(3), 623–645 (2023)
14. Grafakos, L.: Classical Fourier Analysis. Graduate Texts in Mathematics. Springer, New York (2010)
15. Han, R., Wu, S.: A monotone discretization for integral fractional Laplacian on bounded Lipschitz domains: pointwise error estimates under Hölder regularity. SIAM J. Numer. Anal. **60**(6), 3052–3077 (2022)
16. Hernández-Santamaría, V., Jarohs, S., Saldaña, A., Sinsch, L.: FEM for 1D-problems involving the logarithmic Laplacian: error estimates and numerical implementation. Comput. Math. Appl. **192**, 189–211 (2025)
17. Huang, Y., Oberman, A.: Numerical methods for the fractional Laplacian: a finite difference–quadrature approach. SIAM J. Numer. Anal. **52**(6), 3056–3084 (2014)
18. Huang, Y., Oberman, A.: Finite difference methods for fractional Laplacians. arXiv:1611.00164 (2016)
19. Kwaśnicki, M.: Ten equivalent definitions of the fractional Laplace operator. Fract. Calc. Appl. Anal. **20**, 7–51 (2017)
20. Laptev, A., Weth, T.: Spectral properties of the logarithmic Laplacian. Anal. Math. Phys. **11**(3), 1–20 (2021)
21. Qiu, Y.: Spectra: a C + + library for large scale eigenvalue problems. https://spectralib.org/ (2025)
22. Rosenfeld, J.A., Rosenfeld, S.A., Dixon, W.E.: A mesh-free pseudospectral approach to estimating the fractional Laplacian via radial basis functions. J. Comput. Phys. **390**, 306–322 (2019)
23. Striet, L.: Approximation of some nonlocal operators in a sinc-basis. PhD Thesis, University of Freiburg (2024)
24. Valdinoci, E.: From the long jump random walk to the fractional Laplacian. Preprint (2009)
25. Zhuang, Q., Heryudono, A., Zeng, F., Zhang, Z.: Radial basis methods for integral fractional Laplacian using arbitrary radial functions. SSRN Preprint 4283586 (2022)

A Survey on the Resolvent Convergence

Joaquim Duran

Abstract This chapter deals with the notion of the resolvent of a self-adjoint operator. We pay special attention to the convergence of unbounded self-adjoint operators in several resolvent senses, and how they are related to the convergence of their spectra. We also explore the relations that these notions of convergence have with the so-called strong graph limit, G-convergence, and Γ-convergence.

1 Motivation for a Complex Setting

The *resolvent* is a suitable notion for studying the spectra and the convergence of densely defined, linear, unbounded, self-adjoint transformations on infinite dimensional, separable Hilbert spaces. Throughout this chapter we will only consider separable Hilbert spaces over $\mathbb{C}$, which shall be denoted by $(H, \langle\cdot,\cdot\rangle_H, \|\cdot\|_H)$ —or, when no confusion arises, by H—, unless otherwise specified.

Definition 1 An *operator* T on a (complex and separable) Hilbert space H is a linear transformation from its domain, a linear and dense subspace of H denoted by $\mathrm{Dom}(T)$, into H.

Working in the complex setting is not a loss of generality, because every real Hilbert space can be embedded —in the sense that we explain afterward— into a complex Hilbert space —called its *complexification*—, and every real operator can be uniquely extended to a complex operator preserving the embedding —also called its *complexification*. We point out how.

On the one hand, if $(H, \langle\cdot,\cdot\rangle_H, \|\cdot\|_H)$ is a real Hilbert space, then it can be extended to be a complex Hilbert space in the following sense: there exists a Hilbert space $(H^{\mathbb{C}}, \langle\cdot,\cdot\rangle_{H^{\mathbb{C}}}, \|\cdot\|_{H^{\mathbb{C}}})$ over $\mathbb{C}$ and a map $\Upsilon : H \to H^{\mathbb{C}}$ such that:

J. Duran (✉)
Centre de Recerca Matemàtica, Bellaterra, Spain
e-mail: jduran@crm.cat

J. Duran i Lamiel, A. Maione (eds.), *New Frontiers in Homogenization and Fractional Calculus*, Trends in Mathematics,
https://doi.org/10.1007/978-3-032-20098-3_5

1. Υ is linear,
2. $\langle \Upsilon u, \Upsilon v\rangle_{H^{\mathbb{C}}} = \langle u, v\rangle_H$ for all $u, v \in H$, and
3. for every $h \in H^{\mathbb{C}}$ there exist unique $u, v \in H$ such that $h = \Upsilon u + i\Upsilon v$.

Actually, $H^{\mathbb{C}}$ is (up to bijective isometries) the Cartesian product $H^{\mathbb{C}} := H \times H$ endowed with the operations

$$(u_1, v_1)+(u_2, v_2) = (u_1+u_2, v_1+v_2) \quad \text{and} \quad (\alpha+i\beta)(u, v) = (\alpha u-\beta v, \alpha v+\beta u)$$

for $u_1, u_2, v_1, v_2 \in H$ and $\alpha, \beta \in \mathbb{R}$.[1] The scalar product is defined by

$$\langle (u_1, v_1), (u_2, v_2)\rangle_{H^{\mathbb{C}}} := \langle u_1, u_2\rangle_H + \langle v_1, v_2\rangle_H + i\langle v_1, u_2\rangle_H - i\langle u_1, v_2\rangle_H,$$

the associated norm is defined by

$$\|(u, v)\|_{H^{\mathbb{C}}} := \left(\|u\|_H^2 + \|v\|_H^2\right)^{1/2}$$

and the map Υ is the so-called *standard embedding* of H into $H^{\mathbb{C}}$, defined by

$$\begin{aligned} \Upsilon : H &\to H^{\mathbb{C}} \\ u &\mapsto (u, 0). \end{aligned}$$

With an abuse of notation, the complexification of H is sometimes written as $H^{\mathbb{C}} \equiv H \oplus H$, under the identification $(u, v) \equiv u + iv$.

On the other hand, if $T : \mathrm{Dom}(T) \to H$ is an $\mathbb{R}$-linear operator in a real Hilbert space H, then it can be uniquely extended (up to bijective isometries) into a $\mathbb{C}$-linear operator $T^{\mathbb{C}} : \mathrm{Dom}(T)^{\mathbb{C}} \to H^{\mathbb{C}}$ such that $T^{\mathbb{C}}\Upsilon u = \Upsilon T u$ for every $u \in H$. Actually, $T^{\mathbb{C}}$ is the $\mathbb{C}$-linear operator defined by

$$\begin{aligned} T^{\mathbb{C}} : \mathrm{Dom}(T)^{\mathbb{C}} &\to H^{\mathbb{C}} \\ (u, v) &\mapsto (Tu, Tv). \end{aligned}$$

It is clear from the definition that if $\mathrm{Dom}(T)$ is $\mathbb{R}$-linear and dense in H, then $\mathrm{Dom}(T)^{\mathbb{C}}$ is $\mathbb{C}$-linear and dense in $H^{\mathbb{C}}$; hence, $T^{\mathbb{C}}$ is indeed an operator. In addition, it is straightforward to verify that:

1. $T^{\mathbb{C}}$ is self-adjoint in $H^{\mathbb{C}}$ provided that T is self-adjoint in H.
2. $T^{\mathbb{C}}$ is bijective in $H^{\mathbb{C}}$ if and only if T is bijective in H. In such case, $(T^{\mathbb{C}})^{-1} = (T^{-1})^{\mathbb{C}}$.

[1] More generally, this is the definition of the complexification of a real vector space H.

In Definition 2 we recall the notions of *adjoint* of an operator T and of *self-adjointness*.[2] As shall be clear once we define the resolvent of a self-adjoint operator in Sect. 2, if one is given real self-adjoint operators defined on a real Hilbert space, by applying the results of this chapter to the complexifications one can inherit properties of the spectra and the convergence of the original real operators; see Remark 1.5.

Definition 2 Let T be an operator on a Hilbert space H. Let D be the set of $u \in H$ for which there exists $u^* \in H$ satisfying that

$$\langle Tv, u\rangle_H = \langle v, u^*\rangle_H \quad \text{for all } v \in \mathrm{Dom}(T).$$

The *adjoint* of T is the operator whose domain is $\mathrm{Dom}(T^*) := D$, and whose action on each $u \in \mathrm{Dom}(T^*)$ is $T^*u := u^*$. An operator A is called *self-adjoint* if it coincides with its adjoint A^*, namely, if $A = A^*$.

2 Resolvent and Spectrum of an Operator

In the finite dimensional setting, the spectrum of a self-adjoint operator A is the set of its eigenvalues, that is, the numbers $\lambda \in \mathbb{R}$ for which the operator $A - \lambda I$ is not injective; I denotes here the identity. In finite dimensions, this is equivalent to saying that the spectrum is the set of $\lambda \in \mathbb{R}$ for which the operator $A - \lambda I$ is not a bijection.

In the infinite dimensional setting, it is in general no longer true that the set of $\lambda \in \mathbb{R}$ for which the operator $A - \lambda I$ is not a bijection consists of eigenvalues of A, although they all belong to this set. However, it gives a notion for the *spectrum* of A. This is what motivates the definition of the *resolvent set* and the *resolvent*.

Definition 3 Let $A : \mathrm{Dom}(A) \to H$ be a self-adjoint operator. A number $\lambda \in \mathbb{C}$ is said to be in the *resolvent set* of A, denoted $\rho(A)$, if the operator $A - \lambda I :$ $\mathrm{Dom}(A) \to H$ is a bijection with bounded inverse, where I is the identity in H. In such case,

$$R_\lambda(A) := (A - \lambda I)^{-1}$$

is called the *resolvent* of A at λ. The *spectrum* of A is then defined to be the set $\sigma(A) := \mathbb{C} \setminus \rho(A)$.

With an abuse of notation, from now on we will omit the identity I and write $A - \lambda$.

Remark 1 Several remarks about this definition are in order.

---

[2] The reader shall not confuse the notation T^* for the adjoint of an operator T with the complex conjugate of a number $z \in \mathbb{C}$, denoted $\overline{z}$.

1. If the inverse of $A-\lambda$ exists, then it is linear, and by the closed graph Theorem [6, Theorem III.12] it is also bounded. Hence, the boundedness requirement for the inverse in the definition is redundant. However, the resolvent set uses to appear in the literature defined like this; see [6, Definition in pg. 253] or [8, Section 2.4].
2. By the self-adjointness of A, it holds that $\mathbb{C}\setminus\mathbb{R}\subseteq\rho(A)$ [6, Theorem VIII.3] —and consequently $\sigma(A)\subseteq\mathbb{R}$—, hence $\rho(A)$ is never empty. Actually, for every $u\in\mathrm{Dom}(A)$ there holds

$$\|(A-\lambda)u\|_H^2=\langle(A-\lambda)u,(A-\lambda)u\rangle_H=\|(A-\mathrm{Re}\lambda)u\|_H^2+|\mathrm{Im}\lambda|^2\|u\|_H^2.$$

Hence,

$$\|(A-\lambda)^{-1}\|_{H\to H}\leq\frac{1}{|\mathrm{Im}\lambda|}\qquad\text{for every }\lambda\in\mathbb{C}\setminus\mathbb{R}.\tag{1}$$

3. By the self-adjointness of A, the adjoint of the resolvent $R_\lambda(A)$ satisfies $R_\lambda(A)^*=R_{\overline{\lambda}}(A)=(A-\overline{\lambda})^{-1}$. Indeed, since

$$\begin{aligned}\langle u,v\rangle_H&=\langle(A-\lambda)R_\lambda(A)u,v\rangle_H=\langle R_\lambda(A)u,(A-\overline{\lambda})v\rangle_H\\&=\langle u,R_\lambda(A)^*(A-\overline{\lambda})v\rangle_H\end{aligned}$$

holds for every $u\in H$ and every $v\in\mathrm{Dom}(A)$ —and A is densely defined in H—, we must have $R_\lambda(A)^*(A-\overline{\lambda})=I$.
4. The same definition is meaningful for (densely defined) closed operators, not necessarily self-adjoint. The interested reader may look at [6, Section VIII.1] and [8, Section 2.4], where the resolvents are introduced for such operators —beware that in [6] the resolvent of one such operator T at $\lambda\in\rho(T)$ is defined to be $(\lambda I-T)^{-1}$, with opposite sign. However, for technical reasons, self-adjointness has to be assumed as soon as one deals with the convergence of operators through their resolvents. Since this will be the main topic of this chapter, we already pose ourselves into the self-adjoint setting.
5. Let H be a real Hilbert space and let $A:\mathrm{Dom}(A)\to H$ be a real self-adjoint operator. We could analogously define its resolvent set $\rho(A)$ to be the set of $\lambda\in\mathbb{R}$ for which $A-\lambda:\mathrm{Dom}(A)\to H$ is a bijection with bounded inverse, the resolvent of A at one of such λ to be $R_\lambda(A)=(A-\lambda)^{-1}$, and its spectrum to be the set $\sigma(A)=\mathbb{R}\setminus\rho(A)$; see [4, pg. 133]. However, if $H^{\mathbb{C}}$ and $A^{\mathbb{C}}$ denote the complexifications of H and A, respectively, it easily follows that:

   (a) $A^{\mathbb{C}}$ is self-adjoint in $H^{\mathbb{C}}$,
   (b) $\rho(A^{\mathbb{C}})=\rho(A)\cup(\mathbb{C}\setminus\mathbb{R})$,
   (c) $R_\lambda(A^{\mathbb{C}})=(R_\lambda(A))^{\mathbb{C}}$ for every $\lambda\in\rho(A)$, and
   (d) $\sigma(A^{\mathbb{C}})=\sigma(A)$.

   Hence, there is no loss of generality if we work with the complexifications.

The spectrum of a self-adjoint operator can be decomposed in several ways into several subsets, each of them satisfying some properties; see [6, Sections VI.3 and VII.2] and [8, Section 3.3]. In this chapter, we will be interested in the following spectral decomposition.

Definition 4 Let A be a self-adjoint operator.

1. The *point spectrum* of A, denoted $\sigma_{\mathrm{p}}(A)$, is the set of all the eigenvalues of A.
2. The *discrete spectrum* of A, denoted $\sigma_{\mathrm{d}}(A)$, is the subset of $\sigma_{\mathrm{p}}(A)$ consisting of eigenvalues which are isolated in $\sigma(A)$ and have finite multiplicity.
3. The *essential spectrum* of A, denoted $\sigma_{\mathrm{ess}}(A)$, is the set $\sigma_{\mathrm{ess}}(A) = \sigma(A) \setminus \sigma_{\mathrm{d}}(A)$.

In order to emphasize that the spectrum of A is the disjoint union of its discrete and essential spectra, we write $\sigma(A) = \sigma_{\mathrm{d}}(A) \sqcup \sigma_{\mathrm{ess}}(A)$.

As a consequence of Remark 1.2, the resolvent of a self-adjoint operator A always exists for some λ. Moreover, notice that although A could be unbounded and not defined everywhere in H, its resolvent at every $\lambda \in \rho(A)$ is a bounded operator defined everywhere in H. In addition, as the following theorem asserts — a proof of which can be found in [6, Theorem VIII.2]—, the resolvent $R_\lambda(A)$ is well-behaved in $\lambda \in \rho(A)$.

Theorem 1 *Let A be a self-adjoint operator. The resolvent set $\rho(A)$ is a non-empty open subset of $\mathbb{C}$ on which the map $\rho(A) \ni \lambda \mapsto R_\lambda(A)$ is analytic. That is, $R_\lambda(A)$ has an absolutely convergent power series expansion around every $\lambda \in \rho(A)$. Moreover, the so-called* first resolvent formula *holds:*

$$R_\lambda(A) - R_\mu(A) = (\lambda - \mu) R_\lambda(A) R_\mu(A) \quad \textit{for every } \lambda, \mu \in \rho(A).$$

3 Resolvent Convergence

Given a family of unbounded self-adjoint operators $\{A_n\}_{n \in \mathbb{N}}$, we want to come up with a notion of convergence to a limiting unbounded self-adjoint operator A, which provides information about the convergence of the spectra.

A main drawback is that the operators A_n and A need not be defined everywhere in H, nor do they even need to have the same domain. As a consequence, the classical notions of convergence of operators in the weak topology, in the strong topology, or in the operator norm fail to be well defined.

Since it is natural to think that self-adjoint operators are close if certain bounded functions of them are close, resolvents seem to be a convenient tool to provide a notion of convergence. One advantage is that resolvents are bounded operators defined everywhere, and hence the classical notions of convergence are well defined. Another advantage is that, no matter which the family $\{A_n\}_n$ and the limiting operator A are, by self-adjointness A_n and A have resolvents at every $\lambda \in \mathbb{C} \setminus \mathbb{R}$; recall Remark 1.2. All these considerations motivate the following definition.

Definition 5 Let A_n and A be self-adjoint operators. We say that A_n converges to A, as $n \to +\infty$,

1. *in the norm resolvent sense* if $R_\lambda(A_n)$ converges to $R_\lambda(A)$ in the operator norm for every $\lambda \in \mathbb{C} \setminus \mathbb{R}$. In such case, we write $A_n \xrightarrow[n\to+\infty]{\text{n.r.s.}} A$;
2. *in the strong resolvent sense* if $R_\lambda(A_n)u$ converges to $R_\lambda(A)u$ strongly for every $u \in H$ and every $\lambda \in \mathbb{C} \setminus \mathbb{R}$. In such case, we write $A_n \xrightarrow[n\to+\infty]{\text{s.r.s.}} A$.

We shall see in Sect. 4 that these notions of convergence indeed give information about the convergence of the spectra of self-adjoint operators.

Despite the previous considerations, one could expect these definitions to require convergence for every $\lambda \in \mathbb{C} \setminus \big(\sigma(A) \cup \bigcup_{n\in\mathbb{N}} \sigma(A_n)\big)$. The following result, a proof of which can be found in [8, Corollary 6.32], shows that this is actually irrelevant, because resolvent convergence is enough to be tested in a single $\lambda_\star \in \mathbb{C} \setminus \big(\sigma(A) \cup \bigcup_{n\in\mathbb{N}} \sigma(A_n)\big)$.

Theorem 2 *Suppose that A_n converges to A, as $n \to +\infty$, either in the strong or in the norm resolvent sense for one $\lambda_\star \in \mathbb{C} \setminus \left(\sigma(A) \cup \bigcup_{n\in\mathbb{N}} \sigma(A_n)\right)$, namely, that $R_{\lambda_\star}(A_n)$ converges to $R_{\lambda_\star}(A)$ either in operator norm or strongly. Then the same is true for every $\lambda \in \mathbb{C} \setminus \left(\sigma(A) \cup \bigcup_{n\in\mathbb{N}} \sigma(A_n)\right)$.*

As an immediate consequence of this theorem, we can see that the resolvent convergence is invariant under adding real multiples of the identity. Indeed, if A_n and A are self-adjoint operators such that A_n converges to A in the norm (strong) resolvent sense, then also for every $\mu \in \mathbb{R}$ the operators $(A_n + \mu)$ and $(A + \mu)$ are self-adjoint, and $R_i(A_n + \mu) = R_{i-\mu}(A_n)$ converges to $R_i(A + \mu) = R_{i-\mu}(A)$ in the operator norm (strongly). This proves the following result.

Corollary 1 *Let $\mu \in \mathbb{R}$. $(A_n + \mu)$ converges to $(A + \mu)$ in the norm (strong) resolvent sense if and only if A_n converges to A in the norm (strong) resolvent sense.*

We can analogously define the notion of *weak resolvent convergence*.

Definition 6 Let A_n and A be self-adjoint operators. We say that A_n converges to A, as $n \to +\infty$, *in the weak resolvent sense* if $R_\lambda(A_n)u$ converges to $R_\lambda(A)u$ weakly for every $u \in H$ and every $\lambda \in \mathbb{C} \setminus \mathbb{R}$. In such case, we write $A_n \xrightarrow[n\to+\infty]{\text{w.r.s.}} A$.

It is clear that strong resolvent convergence leads to weak resolvent convergence. The following theorem, a proof of which can be found in [8, Lemma 6.37], states that these notions are actually equivalent for self-adjoint operators.

Theorem 3 *Let A_n and A be self-adjoint operators. If $R_{\lambda_\star}(A_n)$ converges to $R_{\lambda_\star}(A)$ weakly in H for some $\lambda_\star \in \mathbb{C} \setminus \mathbb{R}$, then $R_{\lambda_\star}(A_n)$ converges to $R_{\lambda_\star}(A)$ strongly in H.*

It is also clear that norm resolvent convergence leads to strong resolvent convergence. However, the converse is not true in general, as the following example shows.

Example 1 Let $H = L^2(\mathbb{R})$ be the Hilbert space of square integrable, complex valued functions in $\mathbb{R}$, and consider the so-called *position operator* P, defined by

$$\mathrm{Dom}(P) := \left\{ \varphi \in L^2(\mathbb{R}) : \int_{\mathbb{R}} x^2 |\varphi(x)|^2 \, dx < +\infty \right\},$$

$$P(\varphi(x)) := x\varphi(x) \quad \text{for all } f \in \mathrm{Dom}(P).$$

Notice that P is an unbounded operator in $L^2(\mathbb{R})$, and that it is densely defined in $L^2(\mathbb{R})$, because $C_c^\infty(\mathbb{R}) \subset \mathrm{Dom}(P)$. Moreover, it is straightforward to verify that P is self-adjoint; see [8, Example in pg. 68]. We show that the operator $A_n := \frac{1}{n}P$ —de-fined, for $n \in \mathbb{N}$, by $\mathrm{Dom}(A_n) := \mathrm{Dom}(P)$ and $A_n f = \frac{1}{n}Pf$ for every $f \in \mathrm{Dom}(P)$—, converges to the zero operator, as $n \to +\infty$, in the strong resolvent sense but not in the norm resolvent sense.

First, we show the strong resolvent convergence to 0. By Theorem 2, it is enough to show that

$$\|(A_n - i)^{-1} f - (0 - i)^{-1} f\|_{L^2(\mathbb{R})} \to 0 \quad \text{as } n \to +\infty, \text{ for every } f \in L^2(\mathbb{R}).$$

To this end, set $\varphi_n := (A_n - i)^{-1} f$. Then $f = (A_n - i)\varphi_n = (x/n - i)\,\varphi_n$, and hence

$$\begin{aligned} \|(A_n - i)^{-1} f - (0 - i)^{-1} f\|_{L^2(\mathbb{R})}^2 &= \left\| \left(\frac{x}{n} - i\right)^{-1} f - if \right\|_{L^2(\mathbb{R})}^2 \\ &= \int_{\mathbb{R}} \frac{x^2}{x^2 + n^2} |f(x)|^2 \, dx. \end{aligned} \tag{2}$$

Since $\left\{ \mathbb{R} \ni x \mapsto \frac{x^2}{x^2+n^2} |f(x)|^2 \right\}_n$ is a sequence of measurable functions converging pointwise to 0 as $n \to +\infty$, which are all dominated by the integrable function $|f(x)|^2$, the Dominated convergence theorem [8, Theorem A.24] ensures that the integral in (2) converges to 0 as $n \to +\infty$, as desired.

Next, we show that A_n does not converge to 0 in the norm resolvent sense as $n \to +\infty$. To this end, for every $k \in \mathbb{N}$ define $f_k(x) : \mathbb{R} \to \mathbb{C}$ by

$$f_k(x) := \begin{cases} x^{-2} & \text{if } x > k, \\ 0 & \text{elsewhere.} \end{cases}$$

It is clear that $f_k \in L^2(\mathbb{R}) \setminus \{0\}$, and by (2) we have that

$$\frac{\|(A_n - i)^{-1} f_k - (0 - i)^{-1} f_k\|^2_{L^2(\mathbb{R})}}{\|f_k\|^2_{L^2(\mathbb{R})}} = \frac{\int_k^{+\infty} \frac{x^2}{x^2+n^2} |x^{-2}|^2 \, dx}{\int_k^{+\infty} |x^{-2}|^2 \, dx} \geq \frac{k^2}{k^2 + n^2} \quad \text{for all } k, n \in \mathbb{N}.$$

As a consequence,

$$\begin{aligned} &\|(A_n - i)^{-1} - (0 - i)^{-1}\|^2_{L^2(\mathbb{R}) \to L^2(\mathbb{R})} \\ &= \sup_{f \in L^2(\mathbb{R}) \setminus \{0\}} \frac{\|(A_n - i)^{-1} f - (0 - i)^{-1} f\|^2_{L^2(\mathbb{R})}}{\|f\|^2_{L^2(\mathbb{R})}} \\ &\geq \sup_{k \in \mathbb{N}} \frac{k^2}{k^2 + n^2} = 1 \quad \text{for all } n \in \mathbb{N}. \end{aligned}$$

This prevents A_n from converging to 0 in the norm resolvent sense as $n \to +\infty$.

We conclude this section pointing out that, on top of being convenient, these definitions of norm and strong resolvent convergence are good generalizations of the classical convergences in the operator norm and strong topology for bounded operators, respectively.

Theorem 4 *Let A_n and A be uniformly bounded self-adjoint operators. Then, the following statements hold.*

1. *A_n converges to A in the norm resolvent sense if and only if A_n converges to A in operator norm.*
2. *A_n converges to A in the strong resolvent sense if and only if A_n converges to A strongly.*

As a consequence of this theorem, which is a combination of [6, Theorem VIII.18 and Problem VIII.28], the analog for weak resolvent convergence is not true. To see this, let A_n and A be uniformly bounded self-adjoint operators such that A_n converges weakly but not strongly to A, and assume in addition that there was convergence in the weak resolvent sense. Then, by Theorem 3, A_n would converge to A in the strong resolvent sense, and hence, by Theorem 4, A_n would converge to A strongly, leading to a contradiction. We summarize this obstruction in the following corollary.

Corollary 2 *Let A_n and A be uniformly bounded self-adjoint operators. If A_n converges to A weakly but not strongly, then A_n does not converge to A in the weak resolvent sense.*

4 Convergence of Spectra

It is straightforward from Definition 3 that the resolvents of self-adjoint operators encode information about their spectra. In this section, we show how the resolvent convergence actually gives information about the convergence of spectra.

Theorem 5 *Let A_n and A be self-adjoint operators. If A_n converges to A in the strong resolvent sense, then*

$$\sigma(A) \subseteq \lim_{n\to+\infty} \sigma(A_n),$$

meaning that for every $\lambda \in \sigma(A)$ there exists a sequence of $\lambda_n \in \sigma(A_n)$ converging to λ.

This theorem, a proof of which can be found in [6, Theorem VIII.24(a)], says that the spectrum of the limiting operator can not expand if there is strong resolvent convergence. However, in general it can suddenly contract, as the following example shows.

Example 2 We revisit Example 1, where we have shown that, in $H = L^2(\mathbb{R})$, the scaled position operator $A_n := \frac{1}{n}P$ converges to the zero operator in the strong resolvent sense as $n \to +\infty$ (and not in the norm resolvent sense). The limiting operator, $A = 0$, has trivial spectrum $\sigma(A) = \{0\}$. Instead, we show that $\sigma(P) = \mathbb{R}$, and hence $\sigma(A_n) = \mathbb{R}$ for every $n \in \mathbb{N}$.

Let $\lambda \in \mathbb{R}$. In order to see that $\lambda \in \sigma(P)$, we claim that it is enough to show that there exists a sequence $\{\varphi_k\}_{k\in\mathbb{N}} \subset \mathrm{Dom}(P)$ with $\|\varphi_k\|_{L^2(\mathbb{R})} = 1$ satisfying that $\|(P-\lambda)\varphi_k\|_{L^2(\mathbb{R})} \to 0$ as $k \to +\infty$; such a sequence is called a *Weyl sequence*, see [8, Lemma 2.17]. Indeed, if such a sequence exists but λ did not belong to $\sigma(P)$, since in particular λ would be in the resolvent set of P—that is, the resolvent $R_\lambda(P)$ would be a bounded operator—, we could bound

$$1 = \|\varphi_k\|_{L^2(\mathbb{R})} = \|R_\lambda(P)(P-\lambda)\varphi_k\|_{L^2(\mathbb{R})} \leq \|R_\lambda(P)\|_{L^2(\mathbb{R})\to L^2(\mathbb{R})} \|(P-\lambda)\varphi_k\|_{L^2(\mathbb{R})},$$

with the right hand side converging to zero as $k \to +\infty$, reaching a contradiction.

We conclude exhibiting a Weyl sequence for the arbitrary $\lambda \in \mathbb{R}$. For every $k \in \mathbb{N}$, define $\varphi_k : \mathbb{R} \to \mathbb{C}$ as

$$\varphi_k(x) := \begin{cases} \sqrt{k} & \text{if } x \in \left(\lambda - \frac{1}{2k}, \lambda + \frac{1}{2k}\right), \\ 0 & \text{elsewhere.} \end{cases}$$

It is clear that $\varphi_k \in \mathrm{Dom}(P)$ and that $\|\varphi_k\|_{L^2(\mathbb{R})} = 1$ for every $k \in \mathbb{N}$. Moreover,

$$\|(P-\lambda)\varphi_k\|^2_{L^2(\mathbb{R})} = \int_{\lambda-\frac{1}{2k}}^{\lambda+\frac{1}{2k}} (x-\lambda)^2 k \, dx = \frac{1}{12k^2} \longrightarrow 0 \quad \text{as } k \to +\infty.$$

This example shows the possibility that the spectrum of the limiting operator contracts under strong resolvent convergence, by exhibiting some operators that do not, in addition, converge in the norm resolvent sense —recall Example 1. The following theorem, a proof of which can be found in [6, Theorem VIII.23(a)] and [9, Satz 9.24(a)], shows that this can not happen if there is convergence in the norm resolvent sense.

Theorem 6 *Let A_n and A be self-adjoint operators. If A_n converges to A in the norm resolvent sense, then*

$$\sigma(A) = \lim_{n\to+\infty} \sigma(A_n),$$

meaning that:

1. *for every $\lambda \in \sigma(A)$ there exists a sequence of $\lambda_n \in \sigma(A_n)$ converging to λ, and*
2. *if $\lambda_n \in \sigma(A_n)$ converge to some λ, then $\lambda \in \sigma(A)$.*

The same holds true for the essential and discrete spectra, namely,

$$\sigma_{\mathrm{ess}}(A) = \lim_{n\to+\infty} \sigma_{\mathrm{ess}}(A_n) \quad \textit{and} \quad \sigma_{\mathrm{d}}(A) = \lim_{n\to+\infty} \sigma_{\mathrm{d}}(A_n).$$

We conclude this section saying that the convergence of the essential spectra under norm resolvent convergence is not proven throughout [6], but one can find a statement reminiscent of this in [6, Problem VIII.49(c)]. There, it is asserted that if each A_n has purely discrete spectrum in an interval $(a, b) \subset \mathbb{R}$ —meaning that $\sigma(A_n) \cap (a, b) = \sigma_{\mathrm{d}}(A_n) \cap (a, b)$— and A_n converges to A in the norm resolvent sense, then A also has purely discrete spectrum in (a, b).

5 Relation with Strong Graph Limit

In Sect. 3 we have motivated the definition of resolvent convergence, thinking that unbounded self-adjoint operators are close if certain bounded functions of them are close. Alternatively, one could think that unbounded self-adjoint operators are close if their graphs are close. This motivates another notion of convergence which, a priori, seems unrelated to resolvents.

Definition 7 Let A_n be self-adjoint operators in a Hilbert space H. We say that a pair of elements $(\varphi, \psi) \in H \times H$ is in the *strong graph limit* of A_n if there exist $\varphi_n \in \mathrm{Dom}(A_n)$ such that both $\varphi_n \to \varphi$ and $A_n\varphi_n \to \psi$ strongly in H, as $n \to +\infty$. If the set of such pairs is the graph of a self-adjoint operator A, then we say that A is the *strong graph limit* of A_n, and we write $A = \underset{n\to+\infty}{\mathrm{s.gr.lim}}\, A_n$.

Nevertheless, this notion turns out to be equivalent to strong resolvent convergence.

Theorem 7 *Let A_n and A be self-adjoint operators. Then A_n converges to A in the strong resolvent sense if and only if A is the strong graph limit of A_n.*

A proof of this result can be found in [6, Theorem VIII.26]. Similarly, we can define *weak graph limits*.

Definition 8 Let A_n be self-adjoint operators in a Hilbert space H. We say that a pair of elements $(\varphi, \psi) \in H \times H$ is in the *weak graph limit* of A_n if there exist $\varphi_n \in \mathrm{Dom}(A_n)$ such that $\varphi_n \to \varphi$ strongly in H and $A_n\varphi_n \to \psi$ weakly in H, as $n \to +\infty$. If the set of such pairs is the graph of a self-adjoint operator A, then we say that A is the *weak graph limit* of A_n, and we write $A = \underset{n\to+\infty}{\text{w.gr.lim}}\, A_n$.

An analogous statement to Theorem 7 can not hold in the weak sense, because of the following result; see [6, Problem VIII.26].

Theorem 8 *Let A_n and A be uniformly bounded self-adjoint operators. Then A is the weak graph limit of A_n if and only if A_n converges to A in the weak topology.*

Indeed, if an analogous statement to Theorem 7 were true for the weak graph limit, then by Theorem 8 we would have that, for uniformly bounded self-adjoint operators, weak resolvent convergence would be equivalent to convergence in the weak topology, which contradicts Corollary 2. This proves the following statement.

Corollary 3 *Let A_n and A be uniformly bounded self-adjoint operators. If A is the weak graph limit of A_n and A_n does not converge to A strongly, then A_n does not converge to A in the weak resolvent sense.*

6 Relation with G- and Γ-Convergence

We conclude this survey on the resolvent convergence exploring its relation with *G*- and *Γ-convergence*. In the literature —see, for example, [2, 4]—, *G-convergence* of positive self-adjoint operators and *Γ-convergence* of their associated positive quadratic forms is usually formulated in the context of real Hilbert spaces. Nevertheless, many of the proofs developed in the real Hilbert space setting adapt directly to the complex Hilbert space setting. In [1, Section 2] it is described the modifications needed to generalize the pertinent results in [4] to the complex Hilbert space setting. As defined in [1, Definition 1], the notion of Γ-convergence is the same as in [4], and states as follows.

Definition 9 Let $Q_n : H \to \mathbb{R} \cup \{\infty\}$, for $n \in \mathbb{N}$, and $Q_\infty : H \to \mathbb{R} \cup \{\infty\}$ be functionals. We say that Q_n *Γ-converges in the strong topology* to Q_∞, as $n \to +\infty$, if

1. for every $u \in H$ and every u_n converging to u strongly in H, it holds that $Q_\infty(u) \leq \liminf_{n\to+\infty} Q_n(u_n)$, and
2. for every $u \in H$ there exists a sequence u_n converging to u strongly in H such that $Q_\infty(u) = \lim_{n\to+\infty} Q_n(u_n)$.

Remark 2 Two remarks are in order about Definition 9.

1. With an abuse of notation, we alternatively say that Q_n *strongly Γ-converges* to Q_∞, and we write $Q_n \xrightarrow[n\to+\infty]{S\Gamma} Q_\infty$.
2. There exists a notion of "weak" Γ-convergence, the definition of which is not analogous to Definition 9; see [2, equation (1.13) and the paragraph below it]. However, assuming, in addition, that there exists a constant $c > 0$ such that $Q_n(v) \geq c\|v\|_H^2$ for every $v \in H$ and every $n \in \mathbb{N} \cup \{\infty\}$, if in Definition 9 we replace the strong convergence of u_n by weak convergence, then we get a characterization of the so-called *Γ-convergence in the weak topology* of Q_n to Q_∞; see [4, Proposition 8.10]. With an abuse of notation, we alternatively say that Q_n *weakly Γ-converges* to Q_∞, and we write $Q_n \xrightarrow[n\to+\infty]{W\Gamma} Q_\infty$.

Similarly, the notion of G-convergence of equi-coercive self-adjoint operators is the same as in [4, Definition 13.3].

Definition 10 Given $c > 0$, let A_n and A_∞ be self-adjoint operators such that $\langle A_n u, u\rangle_H \geq c\|u\|_H^2$ for every $u \in \mathrm{Dom}(A_n)$ and every $n \in \mathbb{N} \cup \{\infty\}$. We say that A_n *strongly G-converges* to A_∞, as $n \to +\infty$, if for every $f \in H$ the inverse $A_n^{-1} f$ converges to $A_\infty^{-1} f$ strongly in H, and we write $A_n \xrightarrow[n\to+\infty]{SG} A_\infty$. If instead of $A_n^{-1} f$ converging to $A_\infty^{-1} f$ strongly in H one considers weak convergence in H, then we say that A_n *weakly G-converges* to A_∞, and we write $A_n \xrightarrow[n\to+\infty]{WG} A_\infty$.

In view of Theorem 2, if A_n and A_∞ are non-negative self-adjoint operators, then A_n converges to A_∞ in the strong resolvent sense if and only if $A_n + \lambda I$ strongly G-converges to $A_\infty + \lambda I$, for every $\lambda > 0$.

The following theorem relates the resolvent with G- and Γ-convergence. It is the complex version of [4, Theorem 13.5 and Corollary 13.7], a proof of which can be found in [1, Section 2]. In the theorem, the quadratic form associated to a self-adjoint operator A is the form $Q : H \to \mathbb{R} \cup \{\infty\}$ defined by

$$Q(u) := \begin{cases} \langle Au, u\rangle_H & \text{if } u \in \mathrm{Dom}(A), \\ \infty & \text{otherwise.} \end{cases}$$

Theorem 9 *Let A_n and A_∞ be self-adjoint operators such that their associated quadratic forms satisfy $Q_n(u) \geq c\|u\|_H^2$ for every $u \in H$ and every $n \in \mathbb{N} \cup \{\infty\}$, for some $c > 0$. Then, Q_n weakly Γ-converges to Q_∞ if and only if A_n weakly G-converges to A_∞. Moreover, the following statements are equivalent.*

1. *A_n converges in the strong resolvent sense to A_∞.*
2. *A_n strongly G-converges to A_∞.*
3. *Q_n both strongly and weakly Γ-converges to Q_∞.*

We point out that studying the convergence of the quadratic forms associated to self-adjoint operators is of special interest, because by the min-max Theorem [8, Theorem 4.14] the eigenvalues of a self-adjoint operator A which are below its essential spectrum are given by the min-max levels

$$\inf_{\substack{V\subset \mathrm{Dom}(A)\\ \dim(V)=k}} \sup_{u\in V\setminus\{0\}} \frac{\langle Au, u\rangle_H}{\|u\|_H^2}, \quad \text{where } k\in\mathbb{N}.$$

Remark 3 Let $\mu > 0$. In [4, Theorem 13.6] it is shown —in the real Hilbert space setting— the equivalence of the following conditions, assuming that A_n and A_∞ are self-adjoint operators whose associated quadratic forms satisfy $Q_n(u) \geq 0$ for every $u \in H$ and every $n \in \mathbb{N}\cup\{\infty\}$:

1. A_n converges in the strong resolvent sense to A_∞.
2. $(A_n + \mu)$ strongly G-converges to $(A_\infty + \mu)$.
3. $(Q_n + \mu\|\cdot\|_H^2)$ both strongly and weakly Γ-converges to $(Q_\infty + \mu\|\cdot\|_H^2)$.

This also holds in the complex Hilbert space setting. Indeed, by Corollary 1, point 1 is equivalent to $(A_n + \mu)$ converging in the strong resolvent sense to $(A_\infty + \mu)$, and this is in turn equivalent both to points 2 and 3, by Theorem 9.

It is worth noting that, by Theorem 9, the strong resolvent convergence of self-adjoint operators is not equivalent to the strong Γ-convergence of their quadratic forms. The main reason for this is the fact that, in general, the weak and strong Γ-limits of a same sequence may differ; see [4, Example 6.6]. The following result provides a scenario in which the strong resolvent convergence is equivalent to all the notions of convergence that have been presented in this section.

Theorem 10 *Assume that $(h, \langle\cdot,\cdot\rangle_h, \|\cdot\|_h)$ is a compactly embedded Hilbert subspace of H. Let A_n and A_∞ be self-adjoint operators in H such that*

1. $\mathrm{Dom}(A_n) \subseteq h$ *for every $n \in \mathbb{N}\cup\{\infty\}$, and*
2. *their associated quadratic forms are equi-coercive in h, namely, they satisfy*

$$Q_n(u) \geq c\|u\|_h^2 \quad \textit{for every } u \in h \textit{ and every } n \in \mathbb{N}\cup\{\infty\}, \textit{ for some } c > 0.$$

Then, the following statements are equivalent.

1. *A_n converges in the strong resolvent sense to A_∞.*
2. *A_n strongly G-converges to A_∞.*
3. *Q_n strongly Γ-converges to Q_∞.*

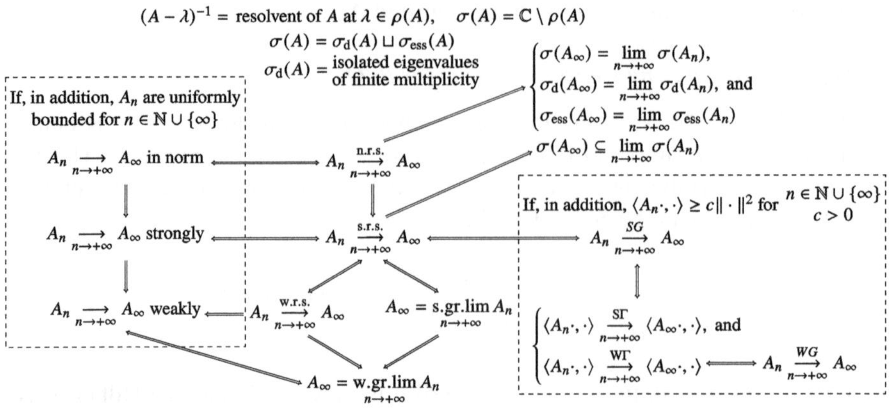

Fig. 1 Diagram of the relations between the different notions of convergence of densely defined, self-adjoint operators A_n, $n \in \mathbb{N} \cup \{\infty\}$, in a separable, complex Hilbert space $(H, \langle\cdot,\cdot\rangle, \|\cdot\|)$

4. *Q_n weakly Γ-converges to Q_∞.*
5. *A_n weakly G-converges to A_∞.*

A proof of this theorem can be found in [4, Theorem 13.12], where the Hilbert spaces are assumed to be real. Nevertheless, the real structure is only used in the proof of [4, Theorem 13.12] to invoke Theorem 9, which also holds in the complex Hilbert space setting.

The scenario described by the hypothesis of Theorem 10 is usually found in the context of uniform elliptic PDEs in divergence form. The reader can find one such concrete example in Sect. 7, where we study the resolvent convergence in $H = L^2(\Omega)$ of the Robin Laplacian to the Dirichlet Laplacian as the Robin boundary condition degenerates to the Dirichlet boundary condition; see the notation at the beginning of Sect. 7. Indeed, the quadratic forms associated to the weak formulations of both the Robin and the Dirichlet Laplacians are equi-coercive in the compactly embedded Hilbert subspace $h = H^1(\Omega)$, on which the operators are defined.

The reader can also find in Fig. 1 a diagram summarizing the results involving the resolvent convergence that we stated in this survey.

7 Example: Resolvent Convergence of the Robin Laplacian

We conclude this chapter illustrating the resolvent convergence through an example. To this end, we first recall the basic notation regarding Sobolev spaces that we will use; see [3, Chapter 9 and Section 11.4], [5, Chapter 5], or [7, Chapter 4].

In the sequel, Ω denotes a bounded domain in $\mathbb{R}^N$ with C^2 boundary, $N \geq 2$. We denote by $L^2(\Omega)$ the Hilbert space of functions $\varphi : \Omega \to \mathbb{C}$ endowed with the scalar product $\langle \cdot, \cdot \rangle_{L^2(\Omega)}$ and the associated norm $\| \cdot \|_{L^2(\Omega)}$, respectively defined as

$$\langle \varphi, \psi \rangle_{L^2(\Omega)} := \int_\Omega \varphi \overline{\psi} \, dx \quad \text{and} \quad \|\varphi\|_{L^2(\Omega)} := \langle \varphi, \varphi \rangle_{L^2(\Omega)}^{1/2}.$$

For $k \in \mathbb{N}$, we denote by $H^k(\Omega)$ the Sobolev space of functions in $L^2(\Omega)$ with weak partial derivatives up to order k in $L^2(\Omega)$, and $H_0^k(\Omega)$ denotes the closure with respect to the $H^k(\Omega)$ norm of the set of smooth functions compactly supported in Ω.

Similarly, $L^2(\partial\Omega)$ denotes the Hilbert space of functions $\varphi : \partial\Omega \to \mathbb{C}$ endowed with the scalar product $\langle \cdot, \cdot \rangle_{L^2(\partial\Omega)}$ and the associated norm $\| \cdot \|_{L^2(\partial\Omega)}$, respectively defined as

$$\langle \varphi, \psi \rangle_{L^2(\partial\Omega)} := \int_{\partial\Omega} \varphi \overline{\psi} \, d\sigma \quad \text{and} \quad \|\varphi\|_{L^2(\partial\Omega)} := \langle \varphi, \varphi \rangle_{L^2(\partial\Omega)}^{1/2}.$$

Here, $\partial\Omega$ denotes the boundary of Ω and σ denotes the surface measure on $\partial\Omega$. We denote by $H^{1/2}(\partial\Omega)$ the fractional Sobolev space of functions $\varphi \in L^2(\partial\Omega)$ such that

$$\|\varphi\|_{H^{1/2}(\partial\Omega)} := \left(\int_{\partial\Omega} |\varphi|^2 \, d\sigma + \int_{\partial\Omega} \int_{\partial\Omega} \frac{|\varphi(x) - \varphi(y)|^2}{|x - y|^2} \, d\sigma(y) \, d\sigma(x) \right)^{1/2} < +\infty.$$

We shall omit the measures of integration when no confusion arises.

The continuous dual space of $H^{1/2}(\partial\Omega)$ is denoted by $H^{-1/2}(\partial\Omega)$. The action of $\varphi \in H^{-1/2}(\partial\Omega)$ on $\psi \in H^{1/2}(\partial\Omega)$ is denoted by the pairing $\langle \varphi, \psi \rangle_{H^{-1/2}(\partial\Omega), H^{1/2}(\partial\Omega)}$, and the norm in $H^{-1/2}(\partial\Omega)$ is

$$\|\varphi\|_{H^{-1/2}(\partial\Omega)} := \sup_{\|\psi\|_{H^{1/2}(\partial\Omega)} \leq 1} \langle \varphi, \psi \rangle_{H^{-1/2}(\partial\Omega), H^{1/2}(\partial\Omega)}.$$

Recall that, whenever $\varphi \in L^2(\partial\Omega) \subset H^{-1/2}(\partial\Omega)$ and $\psi \in H^{1/2}(\partial\Omega) \subset L^2(\partial\Omega)$, the pairing satisfies

$$\langle \varphi, \psi \rangle_{H^{-1/2}(\partial\Omega), H^{1/2}(\partial\Omega)} = \overline{\langle \varphi, \psi \rangle_{L^2(\partial\Omega)}}; \tag{3}$$

see, for example, [3, Remark 3 in Section 5.2, and Section 11.4]. The reason why there is a complex conjugate in (3) is that we defined $\langle \cdot, \cdot \rangle_{L^2(\partial\Omega)}$ to be linear with respect to the first entry.

We want to study the convergence in a resolvent sense of the so-called *Robin Laplacian*, defined for $a \in (0, +\infty)$ by

$$\begin{aligned} \mathrm{Dom}(-\Delta_a) &:= \{u \in H^2(\Omega) : \partial_\nu u + au = 0 \text{ in } H^{1/2}(\partial\Omega)\}, \\ -\Delta_a u &:= -\Delta u \quad \text{for every } u \in \mathrm{Dom}(-\Delta_a), \end{aligned}$$

as the boundary parameter a moves in $(0, +\infty)$. Here, Δ denotes the Laplacian and ν denotes the outward unit normal vector to $\partial\Omega$. Heuristically, the operator formally obtained taking $a = 0$ is the so-called *Neumann Laplacian*, defined by

$$\begin{aligned} \mathrm{Dom}(-\Delta_\mathrm{N}) &:= \{u \in H^2(\Omega) : \partial_\nu u = 0 \text{ in } H^{1/2}(\partial\Omega)\}, \\ -\Delta_\mathrm{N} u &:= -\Delta u \quad \text{for every } u \in \mathrm{Dom}(-\Delta_\mathrm{N}). \end{aligned}$$

Instead, the operator formally obtained taking $a = +\infty$ is the so-called *Dirichlet Laplacian*, defined by

$$\begin{aligned} \mathrm{Dom}(-\Delta_\mathrm{D}) &:= H^2(\Omega) \cap H_0^1(\Omega), \\ -\Delta_\mathrm{D} u &:= -\Delta u \quad \text{for every } u \in \mathrm{Dom}(-\Delta_\mathrm{D}). \end{aligned}$$

As we shall see later, the resolvent convergence formalizes these heuristics.

It is well-known that the above operators are self-adjoint in $L^2(\Omega)$. In particular, by Remark 1.2 the resolvent of $-\Delta_a$ at every $\lambda \in \mathbb{C} \setminus \mathbb{R}$ is a well defined bounded operator from $L^2(\Omega)$ to $L^2(\Omega)$, for every $a \in (0, +\infty)$. One might have the impression that working with the resolvent $(-\Delta_a - \lambda)^{-1}$ could be complicated, because it is the inverse of an operator. However, given $f \in L^2(\Omega)$, the resolvent $(-\Delta_a - \lambda)^{-1} f \in \mathrm{Dom}(-\Delta_a)$ is by definition the (unique) solution φ_a to the elliptic boundary value problem

$$\begin{cases} (-\Delta - \lambda)\varphi_a = f & \text{in } L^2(\Omega), \\ \partial_\nu \varphi_a + a\varphi_a = 0 & \text{in } H^{1/2}(\partial\Omega). \end{cases} \tag{4}$$

Hence, applying the modern theory of elliptic PDEs —see [3, Chapter 9], [5, Chapter 6] or [7, Chapter 5]— we can extract a lot of information from $(-\Delta_a - \lambda)^{-1} f$. The following result is an illustration of this, and shows that the resolvent $(-\Delta_a - \lambda)^{-1}$ is actually a bounded operator from $L^2(\Omega)$ to $H^1(\Omega)$.

Proposition 1 *For every $a > 0$ and $\lambda \in \mathbb{C} \setminus \mathbb{R}$, there exists a constant $C_\lambda > 0$ depending only on λ such that*

$$\|(-\Delta_a - \lambda)^{-1} f\|_{H^1(\Omega)} \leq C_\lambda \|f\|_{L^2(\Omega)}$$

for every $f \in L^2(\Omega)$. As a consequence, $(-\Delta_a - \lambda)^{-1}$ is a compact operator from $L^2(\Omega)$ to $H^1(\Omega)$, and $\sigma(-\Delta_a)$ is purely discrete.

Proof Fix $a > 0$. Let $f \in L^2(\Omega)$, and set $\varphi_a := (-\Delta_a - \lambda)^{-1} f$. Since φ_a solves the boundary value problem (4), by multiplying the PDE by $\overline{\varphi_a}$, integrating by parts, and using the boundary condition $\partial_\nu \varphi_a + a\varphi_a = 0$ that $\varphi_a \in \mathrm{Dom}(-\Delta_a)$ satisfies, we get

$$\begin{aligned}\int_\Omega f\,\overline{\varphi_a} + \lambda \int_\Omega |\varphi_a|^2 &= \int_\Omega -\Delta\varphi_a\,\overline{\varphi_a} = \int_\Omega |\nabla\varphi_a|^2 - \int_{\partial\Omega} \partial_\nu\varphi_a\,\overline{\varphi_a} \\ &= \int_\Omega |\nabla\varphi_a|^2 + a\int_{\partial\Omega} |\varphi_a|^2.\end{aligned}$$

Hence, since $a > 0$, by the triangle inequality, the Cauchy-Schwarz inequality, and the fact that

$$\|\varphi_a\|_{L^2(\Omega)} \le \frac{1}{|\mathrm{Im}\lambda|}\|f\|_{L^2(\Omega)}$$

(see Remark 1.1), we have

$$\begin{aligned}\|\nabla\varphi_a\|^2_{L^2(\Omega,\mathbb{C}^n)} &\le \left|\int_\Omega f\,\overline{\varphi_a} + \lambda\int_\Omega |\varphi_a|^2\right| \le \|f\|_{L^2(\Omega)}\|\varphi_a\|_{L^2(\Omega)} + |\lambda|\|\varphi_a\|^2_{L^2(\Omega)} \\ &\le \left(\frac{1}{|\mathrm{Im}\lambda|} + \frac{|\lambda|}{|\mathrm{Im}\lambda|^2}\right)\|f\|^2_{L^2(\Omega)}.\end{aligned}$$

Therefore,

$$\|\varphi_a\|^2_{H^1(\Omega)} = \|\varphi_a\|^2_{L^2(\Omega)} + \|\nabla\varphi_a\|^2_{L^2(\Omega,\mathbb{C}^n)} \le \frac{1+|\mathrm{Im}\lambda|+|\lambda|}{|\mathrm{Im}\lambda|^2}\|f\|^2_{L^2(\Omega)},$$

as desired. To conclude, since $H^1(\Omega)$ is compactly embedded in $L^2(\Omega)$ —because Ω is bounded, see [3, Theorem 9.16]—, by the former inequality we deduce that $(-\Delta_a - \lambda)^{-1}$ is a compact operator from $L^2(\Omega)$ to $L^2(\Omega)$. This, together with [7, Proposition 8.8 in Appendix A and the paragraph below it], yields the discreteness of $\sigma(-\Delta_a)$. □

We are now in the position to formalize the heuristic idea that the Robin Laplacian converges to the Neumann Laplacian as $a \to 0^+$, and to the Dirichlet Laplacian as $a \to +\infty$. For the latter, we recall that the trace operator $t_{\partial\Omega}$ given by the trace theorem from $H^1(\Omega)$ to $H^{1/2}(\partial\Omega)$ [5, Theorem 5.5.1] extends into a bounded linear operator t_ν from $H(\mathrm{div},\Omega) := \{F \in L^2(\Omega,\mathbb{C}^n) : \mathrm{div}F \in L^2(\Omega)\}$ to $H^{-1/2}(\partial\Omega)$ in the following sense: for all $F \in H^1(\Omega,\mathbb{C}^n)$, it holds that $t_\nu F = \nu\cdot t_{\partial\Omega}F$ almost everywhere on $\partial\Omega$. Moreover,

$$\|\nu\cdot t_{\partial\Omega}F\|_{H^{-1/2}(\partial\Omega)} \le C(\|F\|_{L^2(\Omega,\mathbb{C}^n)} + \|\mathrm{div}F\|_{L^2(\Omega)}), \tag{5}$$

for some constant $C > 0$ depending only on Ω.

Proposition 2 *The Robin Laplacian* $-\Delta_a$ *converges*

1. *to the Neumann Laplacian* $-\Delta_{\mathrm{N}}$, *as* $a \to 0^+$, *in the norm resolvent sense;*
2. *to the Dirichlet Laplacian* $-\Delta_{\mathrm{D}}$, *as* $a \to +\infty$, *in the norm resolvent sense;*
3. *to the Robin Laplacian* $-\Delta_{a_0}$, *as* $a \to a_0$, *in the norm resolvent sense, for every* $a_0 > 0$.

Proof First, we prove the convergence as $a \to +\infty$. By Theorem 2, it is enough to prove that the difference of resolvents at $\lambda = i$, given by

$$W_{a,\mathrm{D}} := (-\Delta_a - i)^{-1} - (-\Delta_{\mathrm{D}} - i)^{-1},$$

converges to zero in operator norm as $a \to +\infty$. To this end, let $f, g \in L^2(\Omega)$ and set

$$\begin{aligned}\varphi_a &:= (-\Delta_a - i)^{-1} f \in \mathrm{Dom}(-\Delta_a) \subset H^2(\Omega),\\ \psi &:= (-\Delta_{\mathrm{D}} + i)^{-1} g \in \mathrm{Dom}(-\Delta_{\mathrm{D}}) \subset H^2(\Omega).\end{aligned}$$

By integrating by parts and by Remark 1.3 applied to $-\Delta_{\mathrm{D}}$, we get

$$\begin{aligned}\langle W_{a,\mathrm{D}} f, g\rangle_{L^2(\Omega)} &= \langle(-\Delta_a - i)^{-1} f - (-\Delta_{\mathrm{D}} - i)^{-1} f, g\rangle_{L^2(\Omega)}\\ &= \langle(-\Delta_a - i)^{-1} f, g\rangle_{L^2(\Omega)} - \langle f, (-\Delta_{\mathrm{D}} + i)^{-1} g\rangle_{L^2(\Omega)}\\ &= \langle\varphi_a, (-\Delta_{\mathrm{D}} + i)\psi\rangle_{L^2(\Omega)} - \langle(-\Delta_a - i)\varphi_a, \psi\rangle_{L^2(\Omega)}\\ &= \langle\varphi_a, -\Delta\psi\rangle_{L^2(\Omega)} - \langle-\Delta\varphi_a, \psi\rangle_{L^2(\Omega)}\\ &= -\langle\varphi_a, \partial_\nu\psi\rangle_{L^2(\partial\Omega)} + \langle\partial_\nu\varphi_a, \psi\rangle_{L^2(\partial\Omega)}\\ &= \frac{1}{a}\langle\partial_\nu\varphi_a, \partial_\nu\psi\rangle_{L^2(\partial\Omega)},\end{aligned} \tag{6}$$

where in the last equality we have used the boundary conditions $\partial_\nu\varphi_a + a\varphi_a = 0$ and $\psi = 0$, that $\varphi_a \in \mathrm{Dom}(-\Delta_a)$ and $\psi \in \mathrm{Dom}(-\Delta_{\mathrm{D}})$ satisfy. As we shall see afterward, and in view of (6), in order to prove the convergence of $W_{a,\mathrm{D}}$ to zero in operator norm it is enough to bound the last pairing uniformly in a. This is what we do next.

Since both $\partial_\nu\varphi_a$ and $\partial_\nu\psi$ belong to $H^{1/2}(\partial\Omega)$, by the property of the pairing (3), the trace theorem from $H^1(\Omega)$ to $H^{1/2}(\partial\Omega)$ applied to $\nabla\psi$, and (5) applied to $F = \nabla\varphi_a$, we get

$$\begin{aligned}|\langle\partial_\nu\varphi_a, \partial_\nu\psi\rangle_{L^2(\partial\Omega)}| &\le \|\partial_\nu\varphi_a\|_{H^{-1/2}(\partial\Omega)}\|\nabla\psi\|_{H^{1/2}(\partial\Omega)}\\ &\le C(\|\nabla\varphi_a\|_{L^2(\Omega,\mathbb{C}^n)} + \|\Delta\varphi_a\|_{L^2(\Omega)})\|\psi\|_{H^2(\Omega)},\end{aligned} \tag{7}$$

for some constant $C > 0$ depending only on Ω. On the one hand, since ψ solves the boundary value problem

$$\begin{cases} -\Delta\psi + i\psi = g & \text{in } L^2(\Omega), \\ \psi = 0 & \text{in } H^{1/2}(\partial\Omega), \end{cases}$$

with $g \in L^2(\Omega)$, by standard regularity theory —see, for example, [5, Theorem 4 in Section 6.3.2 and Remark in pg. 335]— we have

$$\|\psi\|_{H^2(\Omega)} \leq C\|g\|_{L^2(\Omega)}, \tag{8}$$

for some constant $C > 0$ depending only on Ω. On the other hand, since $(-\Delta_a - i)\varphi_a = f$, by the triangle inequality and (1) we have

$$\|\Delta\varphi_a\|_{L^2(\Omega)} = \|f + i\varphi_a\|_{L^2(\Omega)} \leq \|f\|_{L^2(\Omega)} + \|\varphi_a\|_{L^2(\Omega)} \leq 2\|f\|_{L^2(\Omega)},$$

Moreover, by Proposition 1, it holds that $\|\nabla\varphi_a\|_{L^2(\Omega,\mathbb{C}^n)} \leq 3\|f\|_{L^2(\Omega)}$, and hence

$$\|\nabla\varphi_a\|_{L^2(\Omega,\mathbb{C}^n)} + \|\Delta\varphi_a\|_{L^2(\Omega)} \leq 5\|f\|_{L^2(\Omega)}. \tag{9}$$

Combining (7), (8) and (9) with (6), we obtain

$$|\langle W_{a,\mathrm{D}} f, g\rangle_{L^2(\Omega)}| \leq K/a\|f\|_{L^2(\Omega)}\|g\|_{L^2(\Omega)},$$

for some constant $K > 0$ depending only on Ω. Dividing both sides of this last inequality by $\|f\|_{L^2(\Omega)}\|g\|_{L^2(\Omega)}$, and taking the supremum among all functions $f, g \in L^2(\Omega) \setminus \{0\}$, we conclude that

$$\|W_{a,\mathrm{D}}\|_{L^2(\Omega)\to L^2(\Omega)} \leq K/a.$$

Norm resolvent convergence follows by letting $a \to +\infty$.

The proof of the convergence to the Neumann Laplacian as $a \to 0^+$ is analogous. As before, we will show that

$$W_{a,\mathrm{N}} := (-\Delta_a - i)^{-1} - (-\Delta_\mathrm{N} - i)^{-1},$$

converges to zero in operator norm as $a \to 0^+$. Given $f, g \in L^2(\Omega)$, set

$$\varphi_a := (-\Delta_a - i)^{-1} f \in \mathrm{Dom}(-\Delta_a) \subset H^2(\Omega),$$
$$\psi := (-\Delta_\mathrm{N} + i)^{-1} g \in \mathrm{Dom}(-\Delta_\mathrm{N}) \subset H^2(\Omega).$$

Similarly as in (6), integration by parts and Remark 1.3 applied to $-\Delta_\mathrm{N}$ straightforwardly lead to

$$\langle W_{a,\mathrm{N}} f, g\rangle_{L^2(\Omega)} = -\langle \varphi_a, \partial_\nu\psi\rangle_{L^2(\partial\Omega)} + \langle \partial_\nu\varphi_a, \psi\rangle_{L^2(\partial\Omega)}.$$

Using now the boundary conditions $\partial_\nu \varphi_a + a\varphi_a = 0$ and $\partial_\nu \psi = 0$, that $\varphi_a \in \mathrm{Dom}(-\Delta_a)$ and $\psi \in \mathrm{Dom}(-\Delta_{\mathrm{N}})$ satisfy, we get that

$$\langle W_{a,\mathrm{N}} f, g\rangle_{L^2(\Omega)} = -a\langle \varphi_a, \psi\rangle_{L^2(\partial\Omega)}.$$

As we did before, it is enough to bound the pairing in the right hand side uniformly in a. This is what we do next.

Since both φ_a and ψ belong to $H^1(\Omega)$, by the trace theorem from $H^1(\Omega)$ to $H^{1/2}(\partial\Omega)$ we have

$$\begin{aligned}|\langle \varphi_a, \psi\rangle_{L^2(\partial\Omega)}| &\leq \|\varphi_a\|_{L^2(\partial\Omega)}\|\psi\|_{L^2(\partial\Omega)} \leq \|\varphi_a\|_{H^{1/2}(\partial\Omega)}\|\psi\|_{H^{1/2}(\partial\Omega)}\\ &\leq C\|\varphi_a\|_{H^1(\Omega)}\|\psi\|_{H^1(\Omega)},\end{aligned}$$

for some constant $C > 0$ depending only on Ω. On the one hand, by Proposition 1 there holds $\|\varphi_a\|_{H^1(\Omega)} \leq 3\|f\|_{L^2(\Omega)}$. On the other hand, since ψ solves the boundary value problem

$$\begin{cases}-\Delta\psi = g - i\psi & \text{in } L^2(\Omega),\\ \partial_\nu \psi = 0 & \text{in } H^{1/2}(\partial\Omega),\end{cases}$$

then, by the triangle inequality, Cauchy-Schwarz, and (1), it holds that

$$\|\psi\|^2_{H^1(\Omega)} = \|\psi\|^2_{L^2(\Omega)} + \|\nabla\psi\|^2_{L^2(\Omega,\mathbb{C}^n)} \leq \|g\|^2_{L^2(\Omega)} + \left|\int_\Omega (g - i\psi)\,\overline{\psi}\right|^2 \leq 5\|g\|^2_{L^2(\Omega)}.$$

Combining the previous estimates, we obtain the desired uniform bound.

The norm resolvent convergence to $-\Delta_{a_0}$ as $a \to a_0$ readily follows. We sketch the proof omitting the details. Given $f, g \in L^2(\Omega)$, one sets

$$\begin{aligned}\varphi_a &:= (-\Delta_a - i)^{-1} f \in \mathrm{Dom}(-\Delta_a) \subset H^2(\Omega),\\ \varphi_{a_0} &:= (-\Delta_{a_0} + i)^{-1} g \in \mathrm{Dom}(-\Delta_{a_0}) \subset H^2(\Omega)\end{aligned}$$

and straightforwardly verifies that

$$\langle (-\Delta_a - i)^{-1} f - (-\Delta_{a_0} - i)^{-1} f, g\rangle_{L^2(\Omega)} = (a_0 - a)\langle \varphi_a, \varphi_{a_0}\rangle_{L^2(\partial\Omega)}.$$

Applying Proposition 1 to both φ_a and φ_{a_0}, one immediately bounds the pairing in the right hand side uniformly in a and a_0, and concludes the proof of the theorem.

It is clear from the previous proof that the same result remains true if instead of the Laplacian we consider uniformly elliptic operators in divergence form.

Proposition 3 *Let $A = (A_{ij}(x))_{i,j=1,\dots,n} \in C^{0,1}(\Omega, \mathbb{C}^n \times \mathbb{C}^n)$, and assume that A is Hermitian, that is, such that $A_{ij}(x) = \overline{A_{ji}(x)}$ for almost every $x \in \Omega$. Furthermore, assume that there exist $0 < \lambda < \Lambda < +\infty$ such that:*

1. *$|A(x)\xi| \le \Lambda|\xi|$ for every $\xi \in \mathbb{C}^n$, for almost every $x \in \Omega$, and*
2. *$A(x)\xi \cdot \overline{\xi} \ge \lambda|\xi|^2$ for every $\xi \in \mathbb{C}^n$, for almost every $x \in \Omega$.*

Then, the operator defined for $a \in (0, +\infty)$ as

$$\operatorname{Dom}(L_a) := \{u \in H^2(\Omega) : \nu \cdot A\nabla u + au = 0 \textit{ in } H^{1/2}(\partial\Omega)\},$$
$$L_a u := -\operatorname{div}(A\nabla u) \quad \textit{for every } u \in \operatorname{Dom}(L_a),$$

1. *converges, as $a \to +\infty$, in the norm resolvent sense to the operator*

$$\operatorname{Dom}(L_{\mathrm{D}}) := H^2(\Omega) \cap H_0^1(\Omega),$$
$$L_{\mathrm{D}} u := -\operatorname{div}(A\nabla u) \quad \textit{for every } u \in \operatorname{Dom}(L_{\mathrm{D}});$$

2. *converges, as $a \to 0^+$, in the norm resolvent sense to the operator*

$$\operatorname{Dom}(L_{\mathrm{N}}) := \{u \in H^2(\Omega) : \nu \cdot A\nabla u = 0 \textit{ in } H^{1/2}(\partial\Omega)\},$$
$$L_{\mathrm{N}} u := -\operatorname{div}(A\nabla u) \quad \textit{for every } u \in \operatorname{Dom}(L_{\mathrm{N}});$$

3. *converges, as $a \to a_0$, in the norm resolvent sense to L_{a_0}, for every $a_0 > 0$.*

Acknowledgments The author thanks Maicol Caponi, Alessandro Carbotti, Giuseppe Cosma Brusca, and Alberto Maione for enlightening discussions regarding the relations between resolvent and G- and Γ-convergence. The author also thanks the suggestions of the anonymous referee, which have enriched Sect. 6.

Funding information The author is supported by the Spanish grants PID2021-123903NB-I00 and RED2022-134784-T funded by MCIN/AEI/10.13039/501100011033, by ERDF "A way of making Europe", and by the Catalan grant 2021-SGR-00087. This work is supported by the Spanish State Research Agency, through the Severo Ochoa and María de Maeztu Program for Centers and Units of Excellence in R&D (CEX2020-001084-M), and more specifically by the grant CEX2020-001084-M-20-1. The author acknowledges CERCA Programme/Generalitat de Catalunya for institutional support.

References

1. Bedoya, R., de Oliveira, C.R., Verri, A.A.: Complex Γ-convergence and magnetic Dirichlet Laplacian in bounded thin tubes. J. Spectr. Theory **4**, 621–642 (2014)
2. Braides, A.: Γ-Convergence for Beginners. Oxford University Press, New York (2002)

3. Brezis, H.: Functional Analysis, Sobolev Spaces and Partial Differential Equations. Springer, New York (2011)
4. Dal Maso, G.: An Introduction to Γ-Convergence. Birkhäuser, Boston (1993)
5. Evans, L.C.: Partial Differential Equations, 2nd edn. American Mathematical Society, Providence (2010)
6. Reed, M., Simon, B.: Methods of Modern Mathematical Physics. Volume 1: Functional Analysis. Academic Press, London (1980)
7. Taylor, M.E.: Partial Differential Equations I. Basic Theory, 2nd edn. Springer, New York (2011)
8. Teschl, G.: Mathematical Methods in Quantum Mechanics: With Applications to Schrödinger Operators, 2nd edn. American Mathematical Society, Providence (2014)
9. Weidmann, J.: Lineare Operatoren in Hilberträumen. Teil 1 Grundlagen. Teubner Verlag, Wiesbaden (2000)

Fractional Sobolev Spaces via Interpolation, and Applications to Mixed Local–Nonlocal Operators

Alberto Maione

Abstract In this note, we present a well-known connection between the Sobolev-Slobodeckij spaces, also known as *fractional Sobolev spaces*, and *interpolation theory*. We show how fractional Sobolev spaces can be equivalently characterized as real and complex interpolation spaces between Lebesgue spaces and integer-order Sobolev spaces. We also state a spectral theorem for the so-called *mixed local-nonlocal* operators, and show how interpolation theory leads to its proof. This note is intended for early-career researchers, and aims to provide a concise and accessible introduction to the subject.

1 Interpolation Theory

We recall some classical definitions and results from interpolation theory in the setting of *Banach spaces*. We refer the interested reader to [1, 2, 8–10, 12, 13] for a more detailed exposition of the topic.

Definition 1 Let $(X, \|\cdot\|_X)$ and $(Y, \|\cdot\|_Y)$ be Banach spaces. We say that the couple (X, Y) is *compatible* if there exists a Hausdorff topological vector space V such that $X \subseteq V$ and $Y \subseteq V$.

Given a compatible couple of Banach spaces (X, Y), we can define the *sum* and *intersection* spaces as follows

$$X + Y := \{f \in V : f = g + h \text{ with } g \in X \text{ and } h \in Y\} \subseteq V,$$

$$X \cap Y := \{f \in V : f \in X \text{ and } f \in Y\} \subseteq V.$$

A. Maione (✉)
Dipartimento di Matematica, Politecnico di Milano, Campus Leonardo, Edificio 14 "Nave", Milano, Italy
e-mail: alberto.maione@polimi.it

J. Duran i Lamiel, A. Maione (eds.), *New Frontiers in Homogenization and Fractional Calculus*, Trends in Mathematics,
https://doi.org/10.1007/978-3-032-20098-3_6

By definition, it holds that

$$X \cap Y \subseteq X \cup Y \subseteq X + Y.$$

We work with the sum space $X+Y$ instead of the union space $X \cup Y$, since the latter is not a vector space, even though both X and Y are Banach spaces. We associate $X+Y$ and $X \cap Y$ with the graph norms, respectively given by

$$\|f\|_{X+Y} := \inf_{f=g+h, g \in X, h \in Y} \{\|g\|_X + \|h\|_Y\}, \tag{1}$$

$$\|f\|_{X \cap Y} := \max(\|f\|_X, \|f\|_Y).$$

Theorem 1 *Assume that* $(X, \|\cdot\|_X)$ *and* $(Y, \|\cdot\|_Y)$ *are normed spaces. Then, the spaces* $(X+Y, \|\cdot\|_{X+Y})$ *and* $(X \cap Y, \|\cdot\|_{X \cap Y})$ *are normed spaces. Moreover, if* X, Y *are complete, then* $X+Y$ *and* $X \cap Y$ *are complete as well.*

Definition 2 Let (X, Y) be a compatible couple of Banach spaces and let W be a Banach space. We say that W is an *intermediate space* between X and Y if

$$X \cap Y \subseteq W \subseteq X + Y.$$

All inclusions are continuous.

A fundamental example comes from classical functional analysis. Let Ω be an open set of finite measure and let $p \leq r \leq q$. Consider the following Banach spaces

$$X = L^p(\Omega), \quad Y = L^q(\Omega), \quad W = L^r(\Omega).$$

By the nesting property of Lebesgue spaces, $L^q(\Omega) \hookrightarrow L^r(\Omega) \hookrightarrow L^p(\Omega)$ and, up to equivalent norms,

$$X + Y = L^p(\Omega), \quad X \cap Y = L^q(\Omega).$$

Then, $W = L^r(\Omega)$ is an intermediate space between $X = L^p(\Omega)$ and $Y = L^q(\Omega)$, since

$$X \cap Y = L^q(\Omega) \subseteq W = L^r(\Omega) \subseteq L^p(\Omega) = X + Y.$$

The space $L^r(\Omega)$ of the previous example is not only an intermediate space between $L^p(\Omega)$ and $L^q(\Omega)$, but actually an *interpolation space*, according to the following definition. We use the following notation.

- $\mathcal{L}(X)$ is the set of linear and bounded transformations $T : X \to X$.
- $\mathcal{L}(X) \cap \mathcal{L}(Y)$ is the set of $T : X + Y \to X + Y$ such that $T|_X \in \mathcal{L}(X)$ and $T|_Y \in \mathcal{L}(Y)$.

- $\mathcal{L}(X_0, X_1)$ is the set of linear and bounded transformations $T : X_0 \to X_1$.
- $\mathcal{L}(X_0, X_1) \cap \mathcal{L}(Y_0, Y_1)$ is the set of $T : X_0 + Y_0 \to X_1 + Y_1$ such that $T|_{X_0} \in \mathcal{L}(X_0, X_1)$ and $T|_{Y_0} \in \mathcal{L}(Y_0, Y_1)$.

Definition 3 Let (X, Y) be a compatible couple of Banach spaces and let W be an intermediate space between X and Y. We say that W is an *interpolation space* between X and Y if for every $T \in \mathcal{L}(X) \cap \mathcal{L}(Y)$ we have

$$T|_W \in \mathcal{L}(W).$$

The reason why, in the previous example, the space $W = L^r(\Omega)$ is an interpolation space between $X = L^p(\Omega)$ and $Y = L^q(\Omega)$ directly follows from the following lines.

Theorem 2 (M. Riesz, 1926; Thorin, 1938, 1948) *Let $1 \le p_0 \le p_1 \le +\infty$, and let $1 \le q_0 \le q_1 \le +\infty$, with $p_0 \le q_0$ and $p_1 \le q_1$.*[1] *Let $T \in \mathcal{L}(L^{p_0}, L^{q_0}) \cap \mathcal{L}(L^{p_1}, L^{q_1})$. For $s \in (0, 1)$, let $p_s \in [p_0, p_1]$, with $\frac{1}{p_s} = (1-s)\frac{1}{p_1} + s\frac{1}{p_0}$, and let $q_s \in [q_0, q_1]$, with $\frac{1}{q_s} = (1-s)\frac{1}{q_1} + s\frac{1}{q_0}$. Then, $T \in \mathcal{L}(L^{p_s}, L^{q_s})$ and*

$$\|T\|_{\mathcal{L}(L^{p_s}, L^{q_s})} := \sup_{h \in L^{p_s}, h \neq 0} \frac{\|Th\|_{L^{q_s}}}{\|h\|_{L^{p_s}}} \le \|T\|^{1-s}_{\mathcal{L}(L^{p_1}, L^{q_1})} \|T\|^{s}_{\mathcal{L}(L^{p_0}, L^{q_0})}, \tag{2}$$

where

$$\|T\|_{\mathcal{L}(L^{p_0}, L^{q_0})} := \sup_{f \in L^{p_0}, f \neq 0} \frac{\|Tf\|_{L^{q_0}}}{\|f\|_{L^{p_0}}}, \qquad \|T\|_{\mathcal{L}(L^{p_1}, L^{q_1})} := \sup_{f \in L^{p_1}, f \neq 0} \frac{\|Tf\|_{L^{q_1}}}{\|f\|_{L^{p_1}}}.$$

Definition 4 Let (X_0, Y_0) and (X_1, Y_1) be compatible couples of Banach spaces, and let W_0 and W_1 be Banach spaces. We say that (W_0, W_1) is an *interpolation couple* with respect to (X_0, Y_0) and (X_1, Y_1) if:

(*a*) W_0 is an intermediate space between X_0 and Y_0.
(*b*) W_1 is an intermediate space between X_1 and Y_1.
(*c*) For every $T \in \mathcal{L}(X_0, X_1) \cap \mathcal{L}(Y_0, Y_1)$ it holds that $T \in \mathcal{L}(W_0, W_1)$.

Remark 1 Let (W_0, W_1) be an interpolation couple with respect to (X_0, Y_0) and (X_1, Y_1). Then, in general, W_0 is *not* an interpolation space between X_0 and Y_0 (equivalently, W_1 is not an interpolation space between X_1 and Y_1). However, in practical situations, it is convenient to work with couples of spaces, since by letting $X := X_0 = X_1$ and $Y := Y_0 = Y_1$, then $W_0 \equiv W_1 =: W$, and

$$W \text{ becomes an interpolation space between} X \text{ and } Y.$$

[1] As proved by Thorin in 1938, the assumption $p_0 \le q_0$ and $p_1 \le q_1$ may be dropped.

By means of the previous definition, we can reinterpret Theorem 2 as follows. For $1 \le p_0 \le p_1 \le +\infty$ and $1 \le q_0 \le q_1 \le +\infty$, let (L^{p_0}, L^{p_1}) and (L^{q_0}, L^{q_1}) be compatible couples. For $s \in (0,1)$, let $r_i \in [p_i, q_i]$, with

$$\frac{1}{r_i} = (1-s)\frac{1}{q_i} + s\frac{1}{p_i} \quad (i = 0, 1).$$

Then, (L^{r_0}, L^{r_1}) is an interpolation couple between the couples (L^{p_0}, L^{p_1}) and (L^{q_0}, L^{q_1}). Additionally, if $p_0 = q_0 =: p$, $p_1 = q_1 =: q$, and $p_s = q_s =: r$, we get that L^r is an interpolation space between L^p and L^q, and we recover the following interpolation inequality (3), as a consequence of (2).

Theorem 3 (Interpolation Inequality) *Let Ω be a bounded open subset of $\mathbb{R}^n$, let $1 \le p \le q \le +\infty$, and let $f \in L^p(\Omega) \cap L^q(\Omega)$. Then, $f \in L^r(\Omega)$ for every $r \in [p,q]$. Moreover, the following estimate holds*

$$\|f\|_{L^r(\Omega)} \le \|f\|^{1-s}_{L^q(\Omega)} \|f\|^s_{L^p(\Omega)}, \tag{3}$$

for some $s \in [0,1]$ such that $\frac{1}{r} = (1-s)\frac{1}{q} + s\frac{1}{p}$.

Other trivial examples of interpolation couples are as follows. Let (X_0, Y_0) and (X_1, Y_1) be compatible couples of Banach spaces. Denoted by

$$W_0 = X_0 + Y_0 \text{ and } W_1 = X_1 + Y_1, \text{ and}$$
$$W_0 = X_0 \cap Y_0 \text{ and } W_1 = X_1 \cap Y_1,$$

then, the couples (W_0, W_1) are interpolation couples with respect to (X_0, Y_0) and (X_1, Y_1).

Let (X_0, Y_0) and (X_1, Y_1) be two compatible couples of Banach spaces, and consider $\mathcal{L}(X_0 + Y_0, X_1 + Y_1)$ and $\mathcal{L}(X_0 \cap Y_0, X_1 \cap Y_1)$, the sets of linear and bounded transformations

$$T : X_0 + Y_0 \to X_1 + Y_1, \quad T : X_0 \cap Y_0 \to X_1 \cap Y_1.$$

The associated norms $\|T\|_{\mathcal{L}(X_0+Y_0, X_1+Y_1)}$ and $\|T\|_{\mathcal{L}(X_0\cap Y_0, X_1\cap Y_1)}$ both satisfy

$$\|T\|_{\mathcal{L}(X_0+Y_0, X_1+Y_1)} := \sup_{f \in X_0+Y_0, f \ne 0} \frac{\|Tf\|_{X_1+Y_1}}{\|f\|_{X_0+Y_0}} \le \max\left(\|T\|_{\mathcal{L}(X_0,X_1)}, \|T\|_{\mathcal{L}(Y_0,Y_1)}\right)$$

$$\|T\|_{\mathcal{L}(X_0\cap Y_0, X_1\cap Y_1)} := \sup_{f \in X_0\cap Y_0, f \ne 0} \frac{\|Tf\|_{X_1\cap Y_1}}{\|f\|_{X_0\cap Y_0}} \le \max\left(\|T\|_{\mathcal{L}(X_0,X_1)}, \|T\|_{\mathcal{L}(Y_0,Y_1)}\right).$$

Definition 5 Let (X_0, Y_0) and (X_1, Y_1) be compatible couples of Banach spaces, and let (W_0, W_1) be an interpolation couple with respect to (X_0, Y_0) and (X_1, Y_1). We say that the interpolation is *exact* if for every $T \in \mathcal{L}(X_0, X_1) \cap \mathcal{L}(Y_0, Y_1)$, which gives $T \in \mathcal{L}(W_0, W_1)$, it holds that

$$\|T\|_{\mathcal{L}(W_0,W_1)} \leq \max\left(\|T\|_{\mathcal{L}(X_0,X_1)}, \|T\|_{\mathcal{L}(Y_0,Y_1)}\right).$$

This requirement is quite restrictive. A weaker version is the following.

Definition 6 Let (X_0, Y_0) and (X_1, Y_1) be compatible couples of Banach spaces, and let (W_0, W_1) be an interpolation couple with respect to (X_0, Y_0) and (X_1, Y_1). We say that the interpolation is *uniform*, if for every $T \in \mathcal{L}(X_0, X_1) \cap \mathcal{L}(Y_0, Y_1)$, there exists a positive constant C, only depending on X_0, X_1, Y_0, Y_1, such that

$$\|T\|_{\mathcal{L}(W_0,W_1)} \leq C \max\left(\|T\|_{\mathcal{L}(X_0,X_1)}, \|T\|_{\mathcal{L}(Y_0,Y_1)}\right).$$

In the setting of Banach spaces, it is shown in [2, Theorem 2.4.2] that

(W_0, W_1) is an interpolation couple if and only if the interpolation is uniform.

To the best of our knowledge, even in general normed space, there are no known counterexamples. We note that the interpolation in Theorem 2 is exact of power s, according to the following definition.

Definition 7 Let (X_0, Y_0) and (X_1, Y_1) be compatible couples of Banach spaces, and let (W_0, W_1) be an interpolation couple with respect to (X_0, Y_0) and (X_1, Y_1). We say that the interpolation is of *power s*, if for every $T \in \mathcal{L}(X_0, X_1) \cap \mathcal{L}(Y_0, Y_1)$, there exist $s \in (0, 1)$ and a positive constant C, only depending on X_0, X_1, Y_0, Y_1, such that

$$\|T\|_{\mathcal{L}(W_0,W_1)} \leq C \|T\|_{\mathcal{L}(X_0,X_1)}^{1-s} \|T\|_{\mathcal{L}(Y_0,Y_1)}^{s}.$$

If $C = 1$, we say that (W_0, W_1) is also exact (of power s).

We conclude this section with the following result, whose proof can be found e.g. in [2, Theorem 2.5.1].

Theorem 4 (Aronszajn-Gagliardo, 1965) *Let (X_0, Y_0) be a compatible couple of Banach spaces, and let W_0 be an interpolation space with respect to (X_0, Y_0). Then, there exists a compatible couple of Banach spaces (X_1, Y_1), and a Banach space W_1 such that W_1 is an interpolation space with respect to (X_1, Y_1), and (W_0, W_1) is an exact interpolation couple with respect to (X_0, Y_0) and (X_1, Y_1).*

2 Interpolation Methods

2.1 Real Interpolation. The K-Method

There are several methods to construct interpolation spaces. In this section, we briefly recall the construction of *real interpolation spaces* between Banach spaces through the so-called K-method by Peetre. The construction proceeds in three steps.

1. Given (X, Y) a compatible couple of Banach spaces, we define an equivalent norm, with respect to (1), for the Banach space $X + Y$.
2. We define the real interpolation space $(X, Y)_{s,p}$, for $p \in [1, +\infty)$, and $s \in (0, 1)$.
3. We show that $(X, Y)_{s,p}$ in an interpolation space between X and Y.

Step 1 Let (X, Y) be a compatible couple of Banach spaces and fix $x > 0$. We define the Peetre K-functional as

$$\begin{aligned} K : X + Y &\to \mathbb{R}_0^+ \\ f &\mapsto K(x, f, X, Y) := \inf_{f=g+h, g\in X, h\in Y} \Big\{ \|g\|_X + x\|h\|_Y \Big\}. \end{aligned}$$

The K-functional defines a norm equivalent to $\|\cdot\|_{X+Y}$ (indeed, they coincide when $x = 1$).[2] We note that for every $x \neq 1$, the K-functional is *not symmetric* with respect to X and Y, but instead satisfies the relation

$$K(x, f, Y, X) = xK(x^{-1}, f, X, Y) \neq K(x, f, X, Y). \tag{4}$$

Step 2 We first recall the definition of weighted Lebesgue spaces.

Definition 8 Let Ω be an open subset of $\mathbb{R}^n$ and $p \in [1, +\infty)$, and let $\omega : \Omega \to (0, +\infty)$ be a measurable weight, positive a.e. We define the weighted Lebesgue space

$$L^p_\omega(\Omega) := \left\{ f \in L^p(\Omega) : \int_\Omega |f(x)|^p \omega(x)\, \mathrm{d}x < \infty \right\},$$

endowed with the norm

$$\|f\|_{L^p_\omega(\Omega)} := \left(\int_\Omega |f(x)|^p\, \omega(x)\, dx \right)^{1/p}.$$

[2] When clear from the context, we abbreviate $K(x, f, X, Y)$ with $K(x, f)$.

Definition 9 Let $p \in [1, +\infty)$, $s \in (0, 1)$ and let $x > 0$. The real interpolation space $(X, Y)_{s,p}$ is the subspace of $X + Y$ defined as

$$(X, Y)_{s,p} := \left\{ f \in X + Y : \frac{K(x, f, X, Y)}{x^s} \in L^p_{\frac{1}{x}}(\mathbb{R}^+) \right\}.$$

We endow $(X, Y)_{s,p}$ with the norm

$$\begin{aligned} \|f\|_{(X,Y)_{s,p}} &= \left\| \frac{K(x, f, X, Y)}{x^s} \right\|_{L^p_{\frac{1}{x}}(\mathbb{R}^+)} \\ &= \left(\int_0^{+\infty} \left| \frac{K(x, f, X, Y)}{x^s} \right|^p \frac{1}{x}\, dx \right)^{1/p} \\ &= \left(\int_0^{+\infty} \frac{K(x, f, X, Y)^p}{x^{sp+1}}\, dx \right)^{1/p}. \end{aligned}$$

Step 3 Properties of the real interpolation spaces.

For every $p \in [1, +\infty)$ and $s \in (0, 1)$, $\left((X, Y)_{s,p}, \|\cdot\|_{(X,Y)_{s,p}}\right)$ is a Banach space. Moreover, the following symmetry property follows by (4)

$$(X, Y)_{s,p} = (Y, X)_{1-s,p}.$$

Theorem 5 *Let $p_1, p_2 \in [1, +\infty)$ and let $s_1, s_2 \in (0, 1)$.*

- *If $1 \le p_1 \le p_2 \le +\infty$, then for every $s \in (0, 1)$*

$$X \cap Y \subseteq (X, Y)_{s,p_1} \subseteq (X, Y)_{s,p_2} \subseteq X + Y.$$

- *If $Y \subseteq X$ and $s_1 < s_2$ satisfy $s_2 - \frac{1}{p_1} > s_1 - \frac{1}{p_2}$, then*

$$(X, Y)_{s_2,p_1} \subseteq (X, Y)_{s_1,p_2}.$$

In particular, for every $p \in [1, +\infty)$

$$(X, Y)_{s_2,p} \subseteq (X, Y)_{s_1,p}. \tag{5}$$

All the above inclusions are continuous embeddings.

Let $p_1 = p_2 = p$. By Theorem 5, $(X, Y)_{s,p}$ is an intermediate space between X, Y.

Theorem 6 *Let $p \in [1, +\infty)$ and $s \in (0, 1)$. Let (X_0, Y_0) and (X_1, Y_1) be compatible couples of Banach spaces. Let $T \in \mathcal{L}(X_0, X_1) \cap \mathcal{L}(Y_0, Y_1)$. Then T*

extends to a bounded linear operator

$$T \in \mathcal{L}((X_0, Y_0)_{s,p}, (X_1, Y_1)_{s,p}).$$

Moreover, for every $T \in \mathcal{L}(X_0, X_1) \cap \mathcal{L}(Y_0, Y_1)$, *it holds that*

$$\|T\|_{\mathcal{L}((X_0,Y_0)_{s,p},(X_1,Y_1)_{s,p})} \leq \|T\|^{1-s}_{\mathcal{L}(X_0,X_1)} \|T\|^{s}_{\mathcal{L}(Y_0,Y_1)}.$$

Let $X_0 = X_1 =: X$ and $Y_0 = Y_1 =: Y$. By Theorem 6, $(X, Y)_{s,p}$ is an interpolation space between X and Y. The interpolation is exact of power s; see Definition 7.

Corollary 1 *Let* (X, Y) *be a compatible couple of Banach spaces. Then, for every* $p \in [1, +\infty)$ *and* $s \in (0, 1)$ *there exists a constant* c*, depending only on* p *and* s*, such that*

$$\|f\|_{(X,Y)_{s,p}} \leq c\|f\|^{1-s}_{Y} \|f\|^{s}_{X} \quad \text{for every } f \in X \cap Y.$$

2.2 Complex Interpolation

In this section, we briefly recall the construction of *complex interpolation spaces* between Banach spaces.

Step 1 Construction of the complex interpolation spaces.

Let (X, Y) be a compatible couple of Banach spaces. Consider the vector space $\mathcal{F}(X, Y)$ of functions $f : \mathbb{C} \to X + Y$ satisfying:

(a) f is holomorphic (analytic) on the set $S = \{z \in \mathbb{C} : 0 < \mathrm{Re}(z) < 1\}$,
(b) f is continuous and bounded on $\overline{S} = \{z \in \mathbb{C} : 0 \leq \mathrm{Re}(z) \leq 1\}$,
(c) the mapping $t \mapsto f(it)$ is continuous from $\mathbb{R}$ to X,
(d) the mapping $t \mapsto f(1 + it)$ is continuous from $\mathbb{R}$ to Y.

We endow $\mathcal{F}(X, Y)$ with the norm

$$\|f\|_{\mathcal{F}(X,Y)} := \max\left\{ \sup_{t\in\mathbb{R}} \|f(it)\|_X, \sup_{t\in\mathbb{R}} \|f(1 + it)\|_Y \right\}.$$

Then, $(\mathcal{F}(X, Y), \|\cdot\|_{\mathcal{F}(X,Y)})$ is a Banach space.

Definition 10 Let $s \in (0, 1)$. The complex interpolation space $(X, Y)_{[s]}$ is the subspace of $X + Y$ defined as

$$(X, Y)_{[s]} := \{g : \exists f \in \mathcal{F}(X, Y) \text{ such that } g(t) = f(s + it) \text{ for } t \in \mathbb{R}\}.$$

We endow $(X, Y)_{[s]}$ with the norm

$$\|g\|_{(X,Y)_{[s]}} := \inf\left\{\|f\|_{\mathcal{F}(X,Y)} : f \in \mathcal{F}(X, Y), g = f(s + i\cdot)\right\}.$$

Step 2 Properties of the complex interpolation spaces.

For every $s \in (0, 1)$, $\left((X, Y)_{[s]}, \|\cdot\|_{(X,Y)_{[s]}}\right)$ is a Banach space. Moreover, the following symmetry property follows by definition

$$(X, Y)_{[s]} = (Y, X)_{[1-s]}.$$

Theorem 7 *Let $p \in [1, \infty)$ and let $s, s_1, s_2 \in (0, 1)$.*

- *If $Y \subseteq X$ and $s_1 < s_2$, then*

$$(X, Y)_{[s_2]} \subseteq (X, Y)_{[s_1]}.$$

- *For every $s \in (0, 1)$, we have*

$$(X, Y)_{s,1} \subseteq (X, Y)_{[s]} \subseteq (X, Y)_{s,\infty}.$$

- *In general, $(X, Y)_{s,p} \neq (X, Y)_{[s]}$ unless $p = 2$ (or in special cases such as Hilbert couples).*

All the above inclusions are continuous embeddings.

For every $s \in (0, 1)$, the space $(X, Y)_{[s]}$ is an intermediate space between X and Y.

Theorem 8 *Let $p \in [1, +\infty)$ and $s \in (0, 1)$. Let (X_0, Y_0) and (X_1, Y_1) be compatible couples of Banach spaces. Let $T \in \mathcal{L}(X_0, X_1) \cap \mathcal{L}(Y_0, Y_1)$. Then T extends to a bounded linear operator*

$$T \in \mathcal{L}((X_0, Y_0)_{[s]}, (X_1, Y_1)_{[s]}).$$

Moreover, the operator norm satisfies the interpolation estimate

$$\|T\|_{\mathcal{L}((X_0,Y_0)_{[s]},(X_1,Y_1)_{[s]})} \leq \|T\|^{1-s}_{\mathcal{L}(X_0,X_1)}\|T\|^{s}_{\mathcal{L}(Y_0,Y_1)}.$$

Let $X_0 = X_1 =: X$ and $Y_0 = Y_1 =: Y$. By Theorem 8, $(X, Y)_{[s]}$ is an interpolation space between X and Y. The interpolation is exact of power s; see Definition 7.

Corollary 2 *Let (X, Y) be a compatible couple of Banach spaces. For every $s \in (0, 1)$ there exists a constant c, depending only on s, such that*

$$\|f\|_{(X,Y)_{[s]}} \leq c\|f\|^{1-s}_{Y}\|f\|^{s}_{X} \quad \text{for every } f \in X \cap Y.$$

3 Sobolev Spaces via Interpolation

Definition 11 Let Ω be a bounded open subset of $\mathbb{R}^n$, let $p \in [1, +\infty)$ and let $s \in \mathbb{R}^+$.

- We define the Sobolev space $W^{1,p}(\Omega)$ as the space

$$W^{1,p}(\Omega) := \left\{ f \in L^p(\Omega) : \nabla f \in L^p(\Omega; \mathbb{R}^n) \right\}$$

and we endow $W^{1,p}(\Omega)$ with the graph norm

$$\|f\|_{W^{1,p}(\Omega)} := \left(\|f\|^p_{L^p(\Omega)} + \|\nabla f\|^p_{L^p(\Omega;\mathbb{R}^n)} \right)^{\frac{1}{p}}.$$

For $m \in \mathbb{Z}^+$, the Sobolev space $W^{m,p}(\Omega)$ is defined as

$$W^{m,p}(\Omega) := \left\{ f \in L^p(\Omega) : D^\alpha f \in L^p(\Omega) \text{ for every } \alpha \in \mathbb{N}^n \text{ with } |\alpha| \leq m \right\},$$

endowed with the graph norm

$$\|f\|_{W^{m,p}(\Omega)} := \left(\sum_{|\alpha| \leq m} \|D^\alpha f\|^p_{L^p(\Omega)} \right)^{1/p},$$

where $\alpha = (\alpha_1, \ldots, \alpha_n) \in \mathbb{N}^n$, $|\alpha| = \sum_{i=1}^n \alpha_i$, and $D^\alpha f := \frac{\partial^{|\alpha|} f}{\partial x_1^{\alpha_1} \ldots \partial x_n^{\alpha_n}}$.

- If $s \in (0, 1)$, we define the Sobolev-Slobodeckij space $W^{s,p}(\Omega)$ as the space

$$W^{s,p}(\Omega) := \left\{ f \in L^p(\Omega) : \frac{|f(x) - f(y)|}{|x - y|^{\frac{n}{p}+s}} \in L^p(\Omega \times \Omega) \right\}$$

and we endow $W^{s,p}(\Omega)$ with the graph norm

$$\begin{aligned} \|f\|_{W^{s,p}(\Omega)} &= \left(\|f\|^p_{L^p(\Omega)} + \int_\Omega \int_\Omega \frac{|f(x) - f(y)|^p}{|x - y|^{n+sp}} \mathrm{d}x\mathrm{d}y \right)^{\frac{1}{p}} \\ &=: \left(\|f\|^p_{L^p(\Omega)} + [f]^p_{W^{s,p}(\Omega)} \right)^{\frac{1}{p}}. \end{aligned}$$

- If $s > 1$, we denote $s = m + \lambda$, with $m = \lfloor s \rfloor \in \mathbb{Z}$ and $\lambda \in (0, 1)$, and set

$$W^{s,p}(\Omega) := \left\{ f \in W^{m,p}(\Omega) : D^\alpha f \in W^{\lambda,p}(\Omega) \text{ for every } \alpha \in \mathbb{N}^n : |\alpha| = m \right\}.$$

As before, we endow $W^{s,p}(\Omega)$ with the norm[3]

$$\|f\|_{W^{s,p}(\Omega)} = \left(\|f\|^p_{W^{m,p}(\Omega)} + \sum_{|\alpha|=m} \|D^\alpha f\|^p_{W^{\lambda,p}(\Omega)} \right)^{\frac{1}{p}}.$$

- If $s = m \in \mathbb{Z}^+$, then, up to equivalent norms,

$$W^{s,p}(\Omega) = W^{m,p}(\Omega).$$

Integer-order Sobolev spaces $W^{m,p}(\Omega)$ are not stable neither by real nor complex interpolation. In the next result we collect some results showing that fractional Sobolev spaces $W^{s,p}(\Omega)$ can always be obtained through real interpolation of integer-order Sobolev families. The situation is instead more delicate for the complex interpolation of integer-order Sobolev families, whenever $p \neq 2$.[4] We refer the interested reader e.g. to [2, 10] for a proof of the next result, in its most general form.[5]

Theorem 9 *Let Ω be an open and bounded subset of $\mathbb{R}^n$ of class $\mathbf{C}^1$ and let $p \in [1, +\infty)$. The following hold:*

- *For every $s \in (0, 1)$, up to equivalent norms,*

$$W^{s,p}(\Omega) = \left(L^p(\Omega), W^{1,p}(\Omega)\right)_{s,p}, \qquad W^{s,2}(\Omega) = \left(L^2(\Omega), W^{1,2}(\Omega)\right)_{[s]}.$$

If $p \neq 2$, then $W^{s,p}(\Omega) \neq \left(L^p(\Omega), W^{1,p}(\Omega)\right)_{[s]}$.

- *More generally, if $s > 1$ and $s \notin \mathbb{Z}^+$, write $s = m + \lambda$ with $m = \lfloor s \rfloor \in \mathbb{Z}^+$ and $\lambda \in (0, 1)$. Then, up to equivalent norms,*

$$W^{s,p}(\Omega) = \left(W^{m,p}(\Omega), W^{m+1,p}(\Omega)\right)_{\lambda,p},$$
$$W^{s,2}(\Omega) = \left(W^{m,2}(\Omega), W^{m+1,2}(\Omega)\right)_{[\lambda]}.$$

If $p \neq 2$, then $W^{s,p}(\Omega) \neq \left(W^{m,p}(\Omega), W^{m+1,p}(\Omega)\right)_{[\lambda]}$.

---

[3] Alternatively, one may define the norm on $W^{s,p}(\Omega)$ using only the Gagliardo seminorm $[D^\alpha f]_{W^{\lambda,p}(\Omega)}$ instead of the full $W^{\lambda,p}$ norm, in order to avoid redundancy, since the L^p-norms of derivatives of order m are already included in $\|f\|_{W^{m,p}(\Omega)}$.

[4] Given the purpose of this note, the authors have chosen not to introduces Besov spaces and Bessel spaces, which are the correct framework for the link between interpolation and Sobolev spaces. In the particular cases stated in Theorem 9, they coincide with the Sobolev spaces under consideration.

[5] We remind that $L^p(\Omega) = W^{0,p}(\Omega)$.

Remark 2 For $p \neq 2$, complex interpolation of Sobolev spaces gives rise to Bessel potential spaces $H^{s,p}(\Omega)$ rather than fractional Sobolev spaces $W^{s,p}(\Omega)$. Similarly, real interpolation at integer indices yields Besov spaces $B^m_{p,p}(\Omega)$, which strictly contain $W^{m,p}(\Omega)$ whenever $p \neq 2$. These distinctions disappear in the Hilbert case $p = 2$, where $H^{s,2}(\Omega) = W^{s,2}(\Omega)$ and $B^m_{2,2}(\Omega) = W^{m,2}(\Omega)$.

As a consequence of Theorem 5, Theorem 7 and Theorem 9, the following embedding results for integer-order and fractional Sobolev spaces follow directly.

Theorem 10 *Let Ω be an open subset of $\mathbb{R}^n$ and let $p \in [1, +\infty)$. The following continuous embeddings hold.*

- *If $s_1, s_2 \in (0, 1)$ with $s_1 \leq s_2$, then*

$$W^{s_2,p}(\Omega) \subseteq W^{s_1,p}(\Omega).$$

- *If Ω is a Lipschitz domain, then for every $s \in (0, 1)$*

$$W^{1,p}(\Omega) \subseteq W^{s,p}(\Omega).$$

- *If Ω is a Lipschitz domain and $s_1, s_2 > 1$ satisfy $s_1 \leq s_2$, then*

$$W^{s_2,p}(\Omega) \subseteq W^{s_1,p}(\Omega).$$

4 Mixed Local–Nonlocal Operators

In the final part of this survey, we consider the so-called *mixed local-nonlocal operators*. These operators arise from the superposition of a local operator (e.g., the Laplacian $-\Delta u$) and a nonlocal operator (e.g., the fractional Laplacian $(-\Delta)^s u$). In recent years, there has been growing interest in studying such operators, both for the mathematical challenges they pose and for their relevance in various applications.

The literature on the subject is quite extensive; we refer the interested reader, for instance, to [3, 7]. In [5, 6, 11], the authors studied the dependence of mixed local-nonlocal operators on a real parameter (not necessarily positive) α. Let

$$\mathcal{L}_\alpha u := -\Delta u + \alpha(-\Delta)^s u,$$

for $\alpha \in \mathbb{R}$ with no a priori restrictions. We recall that, for fixed $s \in (0, 1)$, the fractional Laplacian $(-\Delta)^s u$ can be defined via the Cauchy principal value as

$$(-\Delta)^s u(x) := C(n, s)\, \text{P.V.} \int_{\mathbb{R}^n} \frac{u(x) - u(y)}{|x - y|^{n+2s}}\, dy.$$

Depending on the value of α, we can distinguish the following cases:

- When $\alpha = 0$, we recover the local Laplacian $-\Delta u$.
- When $\alpha > 0$, the operator $\mathcal{L}_\alpha$ is still positive-definite. It can be interpreted as the infinitesimal generator of a stochastic process involving both Brownian motion and a pure jump Lévy process.
- When $-\frac{1}{C} < \alpha < 0$, where $C > 0$ is the optimal constant of the continuous embedding $W_0^{1,2}(\Omega) \subseteq W^{s,2}(\Omega)$, i.e.

$$[u]^2_{W^{s,2}(\Omega)} := \iint_{\mathbb{R}^{2n}} \frac{|u(x) - u(y)|^2}{|x - y|^{n+2s}} \, dxdy \le C \int_\Omega |\nabla u|^2 \, dx ,$$

  the operator remains coercive in this range as well.
- Finally, when $\alpha \le -\frac{1}{C}$, $\mathcal{L}_\alpha$ is not positive-definite, the bilinear form naturally associated to it does not induce a scalar product, nor a norm, the variational spectrum may exhibit negative eigenvalues and the maximum principles may fail.

Let us show how interpolation theory is useful in such a delicate context. Let

$$\mathbb{H}(\Omega) := \{ u \in W^{1,2}(\mathbb{R}^n) : u = 0 \text{ a.e. on } \mathbb{R}^n \setminus \Omega \}.$$

If Ω is, for instance, of class $\mathbf{C}^1$, we identify the space $\mathbb{H}(\Omega)$ with the space $W_0^{1,2}(\Omega)$ as

$$u \in W_0^{1,2}(\Omega) \iff \tilde{u} \in \mathbb{H}(\Omega), \quad \text{with } \tilde{u} := \begin{cases} u(x), & x \in \Omega, \\ 0, & x \in \mathbb{R}^n \setminus \Omega. \end{cases}$$

From now on, we shall always identify $u \in W_0^{1,2}(\Omega)$ with its zero extension to $\mathbb{R}^n$ $\tilde{u} \in \mathbb{H}(\Omega)$.

We note that the assumption that Ω is of class $\mathbf{C}^1$ can be removed by working with the closure of smooth functions. By the Poincaré inequality, the quantity

$$\|u\|_{\mathbb{H}} := \left(\int_\Omega |\nabla u|^2 \, dx \right)^{1/2}, \quad u \in \mathbb{H}(\Omega),$$

defines a norm that turns $\mathbb{H}(\Omega)$ into a (real) Hilbert space, isometric to $W_0^{1,2}(\Omega)$. The space $\mathbb{H}(\Omega)$ is separable and reflexive, $\mathbf{C}_0^\infty(\Omega)$ is dense in $\mathbb{H}(\Omega)$, and $\mathbb{H}(\Omega)$ compactly embeds into $L^2(\Omega)$ and into

$$W_0^{s,2}(\Omega) := \left\{ u \in W^{s,2}(\mathbb{R}^n) : u = 0 \text{ a.e. on } \mathbb{R}^n \setminus \Omega \right\}.$$

Let us consider the following Dirichlet boundary value problem

$$\begin{cases} \mathcal{L}_\alpha u = \lambda u & \text{in } \Omega, \\ u = 0 & \text{in } \mathbb{R}^n \setminus \Omega. \end{cases} \tag{6}$$

Definition 12 We say that $\lambda \in \mathbb{R}$ is a Dirichlet eigenvalue of $\mathcal{L}_\alpha$ if there exists a weak solution $u \in \mathbb{H}(\Omega)$ of (6) or, equivalently, if

$$\begin{aligned} \mathcal{B}_\alpha(u, \varphi) &:= \int_\Omega \langle \nabla u, \nabla \varphi \rangle \, dx + \alpha \iint_{\mathbb{R}^{2n}} \frac{(u(x) - u(y))(\varphi(x) - \varphi(y))}{|x - y|^{n+2s}} \, dxdy \\ &= \lambda \int_\Omega u\varphi \, dx \end{aligned}$$

for every $\varphi \in \mathbb{H}(\Omega)$. If such function u exists, we call it eigenfunction corresponding to the eigenvalue λ.

The final result of this exposition is a complete description of the (discrete) spectrum of $\mathcal{L}_\alpha$.

Theorem 11 *For every fixed value of $\alpha \in \mathbb{R}$ the following statements hold true:*

(a) The operator $\mathcal{L}_\alpha$ admits a divergent sequence of eigenvalues $\{\lambda_k\}_{k\in\mathbb{N}}$, bounded from below. That is, there exists $C > 0$ such that

$$-C < \lambda_1 \leq \lambda_2 \leq \ldots \leq \lambda_k \to +\infty, \quad \text{as } k \to +\infty.$$

Moreover, for every $k \in \mathbb{N}$, the eigenvalue λ_k admits the variational characterization

$$\begin{aligned} \lambda_k &= \min_{u\in\mathbb{P}_k, \|u\|_{L^2(\Omega)}=1} \left\{ \int_\Omega |\nabla u|^2 \, dx + \alpha \iint_{\mathbb{R}^{2n}} \frac{|u(x) - u(y)|^2}{|x - y|^{n+2s}} \, dxdy \right\} \\ &= \min_{u\in\mathbb{P}_k, \|u\|_{L^2(\Omega)}=1} \mathcal{B}_\alpha(u, u). \end{aligned} \tag{7}$$

Here $\mathbb{P}_1 := \mathbb{H}(\Omega)$, and, for every $k \geq 2$,

$$\mathbb{P}_k := \left\{ u \in \mathbb{H}(\Omega) : \mathcal{B}_\alpha(u, u_j) = 0 \, \text{for every} \, j = 1, \ldots, k-1 \right\},$$

with u_j eigenfunction corresponding to the eigenvalue λ_j.

(b) For every $k \in \mathbb{N}$, there exists an eigenfunction $u_k \in \mathbb{H}(\Omega)$ corresponding to λ_k, which realizes the minimum in (7).

(c) The sequence $\{u_k\}_{k\in\mathbb{N}}$ of eigenfunctions is an orthonormal basis of $L^2(\Omega)$, and orthogonal also with respect to the bilinear form $\mathcal{B}_\alpha$.

(d) Each eigenvalue λ_k has finite multiplicity.

The proof of Theorem 11 can be found in [5, 11] and a sketch of it is reported below. The key point of the proof is the following interpolation inequality, which follows from (5), and whose proof can be found in [4].

Theorem 12 *Let Ω be a Lipschitz bounded domain in $\mathbb{R}^n$. There exists a positive constant C such that*

$$[u]^2_{W^{s,2}(\Omega)} \leq C\|u\|^{2(1-s)}_{L^2(\Omega)} \|u\|^{2s}_{W^{1,2}(\Omega)}$$

for every $u \in \mathbb{H}(\Omega)$.

The optimal constant C is explicitly computed with the help of Fourier transform; see e.g., [5, 6].

Sketch of the Proof of Theorem 11 We limit this brief exposition to demonstrating the existence of eigenvalues for the operator $\mathcal{L}_\alpha$. We distinguish the following scenarios, according with the value of α.

- If $\alpha > -\frac{1}{C}$, $C > 0$ being the constant of the embedding $W^{1,2}_0(\Omega) \subseteq W^{s,2}(\Omega)$, the conclusions follows by the classical spectral theorem.
- In what follows we assume $\alpha \leq -\frac{1}{C}$.

By Theorem 12 and the Young inequality, with exponents $\frac{1}{s}$ and $\frac{1}{1-s}$, for every $\varepsilon \in \mathbb{R}^+$ there exist positive constants c_ε, c_1, c_2, depending only on s and ε, such that

$$\begin{aligned} |\alpha|[u]^2_{W^{s,2}(\Omega)} &\leq |\alpha|C\left((1-s)c_\varepsilon\|u\|^2_{L^2(\Omega)} + s\varepsilon\|u\|^2_{W^{1,2}(\Omega)}\right) \\ &= |\alpha|c_1\varepsilon\|u\|^2_{\mathbb{H}(\Omega)} + |\alpha|c_2\|u\|^2_{L^2(\Omega)} \end{aligned}$$

for every $u \in \mathbb{H}(\Omega)$. By choosing $\varepsilon = \frac{1}{2c_1|\alpha|}$, we get that

$$|\alpha|[u]^2_{W^{s,2}(\Omega)} \leq \frac{1}{2}\|u\|^2_{\mathbb{H}(\Omega)} + \gamma\|u\|^2_{L^2(\Omega)}$$

for every $u \in \mathbb{H}(\Omega)$, where γ only depends on s, ε and α. Therefore,

$$\mathcal{B}_\alpha(u,u) + \gamma\|u\|^2_{L^2(\Omega)} \geq \frac{1}{2}\|u\|^2_{\mathbb{H}(\Omega)} \quad \text{for every } u \in \mathbb{H}(\Omega).$$

Then, the bilinear form $\mathcal{B}^\gamma_\alpha : \mathbb{H}(\Omega) \times \mathbb{H}(\Omega) \to \mathbb{R}$, defined as

$$\mathcal{B}^\gamma_\alpha(u,v) := \mathcal{B}_\alpha(u,v) + \gamma(u,v)_{L^2(\Omega)} \quad \text{for all } u, v \in \mathbb{H}(\Omega),$$

is symmetric, continuous, and coercive.

The associated linear and continuous operator $\mathcal{L}_\alpha^\gamma$ is related with $\mathcal{L}_\alpha$ by

$$\mathcal{L}_\alpha^\gamma u = \mathcal{L}_\alpha u + \gamma u \quad \text{for every } u \in \mathbb{H}(\Omega).$$

By Lax-Milgram Lemma, $\mathcal{L}_\alpha^\gamma$ is invertible and the inverse operator $(\mathcal{L}_\alpha^\gamma)^{-1}$ is linear and continuous. We also note that $\lambda \in \mathbb{R}$ is an eigenvalue of $\mathcal{L}_\alpha$ in $\mathbb{H}(\Omega)$, with eigenvector $u \in \mathbb{H}(\Omega) \setminus \{0\}$, if and only $\frac{1}{\lambda+\gamma}$ is an eigenvalue of $R^\gamma := (\mathcal{L}_\alpha^\gamma)^{-1}$ in $L^2(\Omega)$, with eigenvector $u \in L^2(\Omega) \setminus \{0\}$. The operator $R^\gamma : L^2(\Omega) \to L^2(\Omega)$ is linear, continuous, and compact, since the embedding $\mathbb{H}(\Omega) \subset L^2(\Omega)$ is compact. Moreover, R^γ is injective and self-adjoint, being $\mathcal{B}_\alpha^\gamma$ symmetric, and

$$(R^\gamma h, h)_{L^2(\Omega)} = \mathcal{B}_\alpha^\gamma(R^\gamma h, R^\gamma h) \geq 0 \quad \text{for all } h \in L^2(\Omega).$$

Therefore, there exists a decreasing sequence $(\mu_k)_k$ of eigenvalues of R^γ with $\mu_k > 0$ and $\mu_k \to 0$ as $k \to \infty$. Moreover, every μ_k has finite multiplicity and there exists a orthonormal basis $(e_k)_k$ of $L^2(\Omega)$ given by eigenvectors of R^γ associated to μ_k. Hence, if we consider

$$\lambda_k := \frac{1}{\mu_k} - \gamma \quad \text{for all } k \in \mathbb{N},$$

we get that $(\lambda_k)_k$ is an increasing sequence of eigenvalues of $\mathcal{L}_\alpha$ in $\mathbb{H}(\Omega)$.

The conclusions follow again from the standard proof of the spectral theorem.

□

Acknowledgments The author wishes to thank Augusto Visintin, who introduced him to the topic, and Maicol Caponi, Nicolò Cangiotti, Enrico Valdinoci and Enzo Vitillaro for many useful suggestions. The author also thanks the anonymous referee for the careful reading and for the helpful suggestions that improved the chapter.

Funding Information The author is a member of GNAMPA of the Istituto Nazionale di Alta Matematica (INdAM), and acknowledges the support of the INdAM - GNAMPA 2025 Project "Metodi variazionali per problemi dipendenti da operatori frazionari isotropi e anisotropi" (Grant Code: CUP_E5324001950001). He is also supported by the Spanish State Research Agency, through the Severo Ochoa and María de Maeztu Program for Centers and Units of Excellence in R&D (CEX2020-001084-M), by MCIN/AEI/10.13039/501100011033 (PID2021-123903NB-I00), and by Generalitat de Catalunya (2021-SGR-00087).

Competing Interests The author has no conflicts of interest to declare.

References

1. Aronszajn, N.: Boundary values of functions with finite Dirichlet integral. Technical Report. University of Kansas, 14, pp. 77–94 (1955)
2. Bergh, J., Löfström, J.: Interpolation Spaces. An Introduction. Grundlehren der Mathematischen Wissenschaften, No. 223, Springer-Verlag, Berlin, New York (1976)
3. Biagi, S., Dipierro, S., Valdinoci, E., Vecchi, E.: Mixed local and nonlocal elliptic operators: regularity and maximum principles. Comm. Partial Differ. Equ. **47**(3), 585–629 (2022)
4. Brezis, H., Mironescu, P.: Gagliardo-Nirenberg inequalities and non-inequalities: the full story. Ann. Inst. H. Poincaré C Anal. Non Linéaire **35**(5), 1355–1376 (2018)
5. Cangiotti, N., Caponi, M., Maione, A., Vitillaro, E.: Klein-Gordon-Maxwell equations driven by mixed local-nonlocal operators. Milan J. Math. **91**(2), 375–403 (2023)
6. Cangiotti, N., Caponi, M., Maione, A., Vitillaro, E.: Schrödinger-Maxwell equations driven by mixed local-nonlocal operators. Fract. Calc. Appl. Anal. **27**(2), 677–705 (2024)
7. Dipierro, S., Proietti Lippi, E., Valdinoci, E.: Linear theory for a mixed operator with Neumann conditions. Asymptot. Anal. **128**(4), 571–594 (2022)
8. Gagliardo, E.: Proprietà di alcune classi di funzioni in più variabili. Ricerche Mat. **7**, 102–137 (1958)
9. Leoni, G.: A first course in fractional Sobolev spaces. Grad. Stud. Math. **229**, xv, 586 p. (2023)
10. Lunardi, A.: Interpolation theory. Appunti. Scuola Normale Superiore di Pisa (Nuova Serie) [Lecture Notes. Scuola Normale Superiore di Pisa (New Series)], vol. 16, 3rd edn. Edizioni della Normale, Pisa (2018)
11. Maione A., Mugnai D., Vecchi E.: Variational methods for nonpositive mixed local-nonlocal operators. Fract. Calc. Appl. Anal. **26**(3), 943–961 (2023)
12. Slobodeckiĭ, L. N.: Generalized Sobolev spaces and their application to boundary problems for partial differential equations. Leningrad. Gos. Ped. Inst. Učen. Zap. **197**, 54–112 (1958)
13. Tartar, L.: An introduction to Sobolev spaces and interpolation spaces. Lecture Notes of the Unione Matematica Italiana, vol. 3, Springer, Berlin; UMI, Bologna (2007)

Some Results About Strongly Degenerate Parabolic Equations and Forward–Backward Parabolic Equations

Fabio Paronetto

Abstract Some results are presented for equations of the type $\rho(x,t)u_t + \mathcal{A}u = f$ and $(\rho(x,t)u)_t + \mathcal{A}u = f$, where ρ is a bounded function valued in $\mathbf{R}$, i.e. it can assume positive, zero, and negative values, and $\mathcal{A}$ an elliptic operator.

Starting from existence results, we briefly present some regularity results and a compactness result with respect to G-convergence.

1 Introduction

Consider $p \geqslant 2$, $T > 0$, Ω an open, connected and bounded subset of $\mathbf{R}^n$ and denote by Q the set $\Omega \times (0,T)$. Consider some Carathéodory functions with $a, b, c \in L^\infty(\Omega \times (0,T) \times \mathbf{R} \times \mathbf{R}^n)$ satisfying

$$\begin{cases} \big(a(x,t,u,\xi),\xi\big) \geqslant \lambda|\xi|^p\,, \\ \big|a(x,t,u,\xi)\big| \leqslant \Lambda|\xi|^{p-1}\,, \\ \big|b(x,t,u,\xi)\big| \leqslant L|\xi|^{p-1}\,, \\ \big|c(x,t,u)\big| \leqslant M|u|^{p-1}\,. \end{cases} \tag{1}$$

Consider moreover a function

$$\rho : \Omega \times (0,T) \to \mathbf{R}, \qquad \rho \in L^\infty(\Omega \times (0,T)). \tag{2}$$

F. Paronetto (✉)
Department of Mathematics "Tullio Levi-Civita", Università di Padova, Padua, Italy
e-mail: fabio.paronetto@unipd.it

J. Duran i Lamiel, A. Maione (eds.), *New Frontiers in Homogenization and Fractional Calculus*, Trends in Mathematics,
https://doi.org/10.1007/978-3-032-20098-3_7

The purpose of this note is to present some results about equations of the type

$$\rho(x,t)u_t + \mathcal{A}u = f \qquad \text{in } \Omega \times (0,T)\,, \tag{3}$$

$$\big(\rho(x,t)u\big)_t + \mathcal{A}u = f \qquad \text{in } \Omega \times (0,T)\,. \tag{4}$$

where

$$\mathcal{A}u := -\mathrm{div}\big(a(x,t,u,Du)\big) + b(x,t,u,Du) + c(x,t,u) \tag{5}$$

Notice that ρ is not necessarily positive.

Let us now try to explain the difficulties one may encounter when studying equations like (3) and (4).

- Equations like (3) and (4) have been widely studied with $\rho = \rho(x)$ positive, and clearly in this case the two equations coincide. As long as one considers $\rho(x) \geqslant a > 0$ in Ω, classical techniques for standard parabolic equations (i.e. with $\rho \equiv 1$) to prove existence or to study regularity of the solutions may simply adapted to equations with ρ, while considering $\rho(x) > 0$ *almost everywhere* in Ω, i.e. ρ may degenerate to zero in subsets of measure zero, forces the use of weighted Sobolev spaces and weighted inequalities (see, at this regard, e.g. [6, 9, 10]).

  This is due to the fact that a part of the natural energy is given by (for a fixed $t \in (0,T)$)

$$\int_\Omega u^2(x,t)\rho(x)\,dx\,.$$

- A further difficulty in proving existence of a solution of Eqs. (3) and (4) is given by the possible dependence of ρ also on t, unless ρ admits a partial derivative with respect to t, i.e., for instance, $\rho, \partial_t\rho \in L^\infty(\Omega \times (0,T))$.

  If we only have $\rho \in L^\infty(\Omega \times (0,T))$, i.e. ρ does not admit a weak partial derivative with respect to t, for both Eqs. (3) and (4) a notion of derivative with respect to t is needed, but this notion is weaker than requiring $\partial_t\rho \in L^\infty(\Omega \times (0,T))$ (see Sect. 2).
- When $\rho \geqslant 0$ the equations become strongly degenerate parabolic equations (in fact, it can be seen as a family of elliptic equations in the parameter t in the set $\{\rho = 0\}$) and classical tools for parabolic equations do not work and have to be adapted. Sometimes this can be done adding a small parameter $\epsilon > 0$ to ρ and then trying to let ϵ go to 0. For example, existence of a solution to Eqs. (3) and (4) could be done also in this way, in other cases taking the limit for $\epsilon \to 0^+$ in the proofs is very delicate, as in the regularity results (see, e.g., the results in Sect. 6).
- The difficulties clearly become greater when ρ changes sign, i.e. it assumes both positive and negative values. In this case approximation as in the previous case, i.e. adding $\epsilon > 0$, is not helpful. Just to give another example, compactness

results are not trivial since the natural energy

$$[0, T] \ni t \mapsto \int_\Omega u^2(x,t)\rho(x,t)\,dx \tag{6}$$

in this case is *not* a non-negative function. In this regard see Sects. 3 and 8.

This type of equations seem to be interesting in many areas and have arisen in connection with many different problems. Many diffusion problems lead to mixed problems, partially elliptic and partially parabolic. This situation corresponds to take $\rho \geqslant 0$ in the equations above. This problem is not new: perhaps the first paper about it is [29] (see also [30], Section III.3, [31], Chap. 5). We also refer to [4], chap. 3, and the references therein for many applications.

Also the more general case, i.e. when ρ is both positive and negative, seems to arise in many different problems: in the study of some stochastic differential equation, in the kinetic theory (see, e.g., [5]), in some physical models (like electron scattering and neutron transport).

The first papers, at least to our knowledge, treating forward-backward parabolic equations are: [1] (for this see also Section 3.2.6 in [12]), where the authors considered the following equation

$$x\frac{\partial u}{\partial t}(x,t) - \frac{\partial^2 u}{\partial x^2}(x,t) = f(x,t) \qquad x \in \mathbf{R}\,; \tag{7}$$

later in [14] and [13] the following problem is considered

$$\operatorname{sgn}(x)\frac{\partial u}{\partial t}(x,t) - \frac{\partial^2 u}{\partial x^2}(x,t) + k\,u(x,t) = f(x,t) \qquad x \in \mathbf{R}\,;$$

then, for other applications, we also refer to the references contained in the already quoted paper [14] and in [2] and [3] where Beals treated equations like (3), but always with simple ρ like

$$\rho(x) = x^m,\ m \text{ odd} \qquad \text{or} \qquad \rho(x) = \operatorname{sgn}(x)\,|x|^p\,.$$

Probably the first general paper about it, but with ρ independent of time, is [15].

Among papers taking into account equations with some more general ρ (and $\mathcal{A}$ linear) we recall [28] and [8]. In particular in [8] a coefficient depending also on time is considered and a condition of regularity in time is requested, i.e.

$$\rho = \rho(x,t)\,, \qquad \rho, \rho_t \in L^\infty\,.$$

A more recent paper where the author consider $\rho = \rho(x,t)$ (for linear operators $\mathcal{A}$) is [7]. Another paper we want to recall is [15] which is more general, even if incomplete.

Among quite recent papers, we also mention [11].

Turning to the content of the article, here we briefly list the content of next sections. In Sect. 2 we present some existence result for Eqs. (3) and (4), recalling that the results contained in [18] and [25] are in fact much more general that those presented here.

In Sect. 3 we present an interesting result, that could appear not natural, regarding the function defined in (6).

In Sect. 4 we give a regularity result with respect to time.

In Sect. 5 we give an approximation result via elliptic equations.

In Sects. 6 and 7 we briefly present some regularity results: local boundendness, Harnack inequality and Hölder continuity, but here we confine to consider $\rho \geqslant 0$ or $\rho > 0$.

Finally in Sect. 8 a general result of G-compactness, that prompted us to study these kind of equations.

For the proofs we will refer to the already existing literature.

2 Existence Results

For the definition in the first part of the present section we refer to [18] and to [25].

In the following we consider Ω a bounded open subset of $\mathbf{R}^n$, $T > 0$, $p \geqslant 2$.

Given ρ as in (2) we will often write $\rho(t)$ to denote the function

$$\Omega \ni x \mapsto \rho(x, t).$$

Notice that, since $\rho \in L^\infty(\Omega \times (0, T))$, by Fubini's theorem we have that

$$\rho(t) \in L^\infty(\Omega) \qquad \text{for a.e. } t \in (0, T).$$

We will denote for the sake of simplicity by V the space $W_0^{1,p}(\Omega)$ and by H the space $L^2(\Omega)$, by V' the space $W^{-1,p'}(\Omega)$. Then we have $V \subset H \subset V'$. Similarly we have

$$\mathcal{V} \subset \mathcal{H} \subset \mathcal{V}',$$

where

$$\mathcal{V} := L^p(0, T; W_0^{1,p}(\Omega)), \quad \mathcal{H} = L^2(0, T; L^2(\Omega)), \quad \mathcal{V}' = L^{p'}(0, T; W^{-1,p'}(\Omega)).$$

Definition 1 A function $\rho \in L^\infty(\Omega \times (0, T))$ is said to belong to the class $\mathcal{E}(C_1, C_2)$, $C_1, C_2 \geqslant 0$, if

$$\|\rho\|_\infty \leqslant C_1$$

and it satisfies conditions (i) and (ii) for every $u, v \in W^{1,p}(\Omega)$

$$(i)\ \ t \mapsto \int_\Omega u(x)v(x)\rho(x,t)\,dx \text{ is absolutely continuous in } [0,T]\,,$$

$$(ii)\ \left|\frac{d}{dt}\int_\Omega u(x)v(x)\rho(x,t)\,dx\right| \leqslant C_2\,\|u\|_{W^{1,p}(\Omega)}\|v\|_{W^{1,p}(\Omega)} \qquad \text{for a.e. } t \in [0,T]\,.$$

Remark 1 (Comments About the Class $\mathcal{E}$) Given $\rho \in \mathcal{E}(C_1, C_2)$ one can introduce a family of operators depending on the parameter $t \in [0,T]$, $R(t) : H \to H$ defined as $R(t)u := u\,\rho(t)$, and $\mathcal{R} : \mathcal{H} \to \mathcal{H}$ defined as

$$\mathcal{R}u := u\,\rho. \tag{8}$$

The introduction of the operator $\mathcal{R}$ may seem a complication, but in reality this is only apparent. Indeed the assumptions about ρ enable us to introduce a family of operators we will denote by $\mathcal{R}'$, as we are going to explain, needed in the sequel. One can indeed define for $\rho \in \mathcal{E}(C_1, C_2)$

$$\langle R'(t)u, v\rangle_{V'\times V} := \frac{d}{dt}\int_\Omega u(x)v(x)\rho(x,t)\,dx\,, \qquad u, v \in V,$$

for almost every $t \in [0,T]$. Since the derivative of an absolute continuous function is measurable and, since $\rho \in \mathcal{E}(C_1, C_2)$, the previous derivative turns out to be bounded, we have that

$$[0,T] \ni t \mapsto \langle R'(t)u, v\rangle_{V'\times V} \in L^\infty(0,T).$$

Then we can also define $\mathcal{R}' : \mathcal{V} \to \mathcal{V}'$ as

$$\langle \mathcal{R}'u, v\rangle_{\mathcal{V}'\times\mathcal{V}} := \int_0^T \langle R'(t)u, v\rangle_{V'\times V} dt, \qquad u, v \in \mathcal{V}. \tag{9}$$

Notice that the operator $\mathcal{R}'$ turns out to be linear, self-adjoint and bounded by the constant C_2. Clearly if ρ is differentiable with respect to t and

$$\rho, \frac{\partial \rho}{\partial t} \in L^\infty(\Omega \times (0,T))$$

we have that, for $u, v \in W^{1,p}(\Omega)$

$$\frac{d}{dt}\int_\Omega u(x)v(x)\rho(x,t)\,dx = \int_\Omega u(x)v(x)\frac{\partial \rho}{\partial t}(x,t)\,dx\,.$$

Definition 2 Given $\rho \in \mathcal{E}(C_1, C_2)$, we will write $\mathcal{R} \in \mathcal{E}(C_1, C_2)$ to denote the operator defined in (8), which admits a derivative in the sense defined in (9).

Fig. 1 A simple example of a possible interface Γ

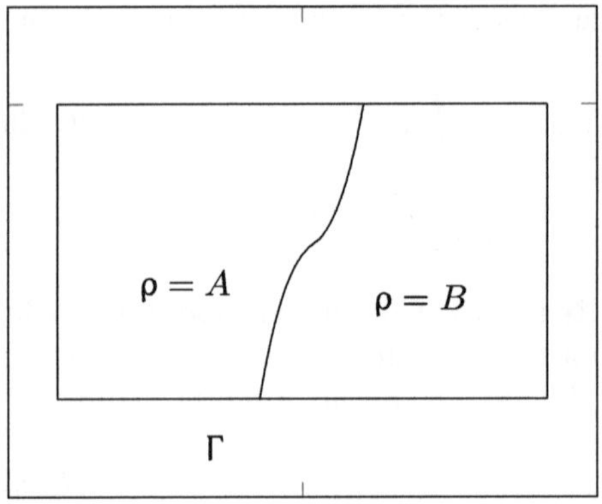

Example of an Admissible ρ Here we give an example of an operator $\mathcal{R} \in \mathcal{E}(C_1, C_2)$ defined via a function $\rho \in \mathcal{E}(C_1, C_2)$ that does not admit a partial derivative with respect to t. Consider $A, B \in \mathbf{R}$ and a partition of $Q = \Omega \times (0, T)$ made by three sets:

$$Q = Q_A \cup \Gamma \cup Q_B$$

where Q_A and Q_B are two open, and for the sake of simplicity connected, sets of Q and $\Gamma = \partial Q_A \cap \partial Q_B$ is a sufficiently regular n-dimensional hypersurface of $\mathbf{R}^{n+1}$, as shown in Fig. 1 below. We will suppose that ρ takes only two values,

$$\rho = A \quad \text{in } Q_A, \qquad \rho = B \quad \text{in } Q_B.$$

Consider now $n = 1$ and $\Omega = (0, L)$. In order to have $\rho \in \mathcal{E}$ we need to suppose that Γ is a graph, precisely suppose Γ is the graph of a function $\gamma : (0, T) \to \mathbf{R}$, differentiable with bounded γ'.

Denote

$$\Omega_A(t) = \{x \in \Omega \mid \rho(x, t) = A\}, \qquad \Omega_B(t) = \{x \in \Omega \mid \rho(x, t) = B\}.$$

We have

$$\Omega_A(t) = (0, \gamma(t)), \qquad \Omega_B(t) = (\gamma(t), L),$$

and

$$(R(t)u, v)_{L^2(0,L)} = \int_0^{\gamma(t)} u(x)v(x)\, A\, dx + \int_{\gamma(t)}^{L} u(x)v(x)\, B\, dx$$

and, for $u, v \in H^1(0, L)$,

$$\left|\frac{d}{dt}(R(t)u, v)_{L^2(0,L)}\right| = \left|A\,\gamma'(t)u(\gamma(t))v(\gamma(t)) - B\,\gamma'(t)u(\gamma(t))v(\gamma(t))\right| \leqslant$$
$$\leqslant |A - B| \|\gamma'\|_\infty |u(\gamma(t))v(\gamma(t))|$$
$$\leqslant C_3 \, \|u\|_{H^1(0,L)} \|v\|_{H^1(0,L)}$$

where $C_3 = C\,|A - B| \|\gamma'\|_\infty$ and C is such that

$$|w(x)| \leqslant C\,\|w\|_{H^1(0,L)} \qquad \text{for every } w \in H^1(0, L) \text{ and for every } x \in [0, L]. \tag{10}$$

Notice that A and B may be positive, null or negative and if $A = B$ the above derivative is zero. An analogous thing holds also in higher dimension.

Moreover, for $u, v \in \mathcal{V}$, we have that

$$\langle \mathcal{R}'u, v\rangle_{\mathcal{V}'\times\mathcal{V}} = \int_0^T u(\gamma(t), t)v(\gamma(t), t)\,dt\,. \tag{11}$$

For many other examples we refer to the last section in [18], to the last section in [25] and to an example in [22] (p. 179).

Notation When we write the product by the function ρ we use the subscript t to denote the temporal derivative, i.e. we write $(\rho u)_t$ and ρu_t. When we use the operator $\mathcal{R}$ we write the superscript $'$ to denote the first derivative (with respect to t), since we look at the equation as an abstract first order differential equation. Therefore we will write $(\mathcal{R}u)'$ for $(\rho u)_t$ and $\mathcal{R}u'$ for ρu_t, finally $\mathcal{R}'$ to denote the operator defined in (9).

For a function $u \in \mathcal{V}$ for which $\mathcal{R}u$ admits generalized derivative in $\mathcal{V}'$ we define the space

$$\mathcal{W}_1 := \left\{u \in \mathcal{V} \mid (\mathcal{R}u)' \in \mathcal{V}'\right\}, \qquad \text{endowed with the norm } \|u\|_{\mathcal{W}_1} := \|u\|_{\mathcal{V}} + \|(\mathcal{R}u)'\|_{\mathcal{V}'}\,. \tag{12}$$

For a function $u \in \mathcal{W}_1$ one can give also a meaning to $\mathcal{R}u'$, that a priori is not defined, as

$$\mathcal{R}u' := (\mathcal{R}u)' - \mathcal{R}'u\,. \tag{13}$$

Then one can also consider the space

$$\mathcal{W}_2 := \{u \in \mathcal{V} \mid \mathcal{R}u' \in \mathcal{V}'\}, \qquad \text{endowed with the norm } \|u\|_{\mathcal{W}_2} := \|u\|_{\mathcal{V}} + \|\mathcal{R}u'\|_{\mathcal{V}'}. \tag{14}$$

One can prove (see Remark 3.1 in [25]) that the two spaces $\mathcal{W}_1$ and $\mathcal{W}_2$ are in fact the same space, which we will denote by $\mathcal{W}$, i.e.

$$\mathcal{W}_1 = \mathcal{W}_2 = \mathcal{W}, \tag{15}$$

and the two norms $\|\cdot\|_{\mathcal{W}_1}$ and $\|\cdot\|_{\mathcal{W}_2}$ are equivalent, provided $\rho \in \mathcal{E}(C_1, C_2)$.

We recall that the operators $\mathcal{R}$ and $\mathcal{R}'$ are defined in Definition 2.

Proposition 1 *If $\rho \in \mathcal{E}(C_1, C_2)$ we have that for every $u, v \in \mathcal{W}$ the function*

$$t \mapsto \int_\Omega u(x,t)v(x,t)\rho(x,t)\,dx$$

is absolutely continuous and there exists a constant c, which depends only on T, such that

$$\begin{aligned}\max_{[0,T]}\left|\int_\Omega u(x,t)v(x,t)\rho(x,t)\,dx\right| \leqslant c\Big[&\|\mathcal{R}u'\|_{\mathcal{V}'}\|v\|_{\mathcal{V}} + \|\mathcal{R}v'\|_{\mathcal{V}'}\|u\|_{\mathcal{V}} + \\ &+ \|\mathcal{R}'\|_{\mathcal{L}(\mathcal{V},\mathcal{V}')}\|u\|_{\mathcal{V}}\|v\|_{\mathcal{V}} + \\ &+ \|\mathcal{R}\|_{\mathcal{L}(\mathcal{H})}\|u\|_{\mathcal{H}}\|v\|_{\mathcal{H}}\Big].\end{aligned}$$

In particular, if $u = v$, we have

$$\max_{[0,T]}\left|\int_\Omega u^2(x,t)\rho(x,t)\,dx\right| \leqslant \tilde{c}\,\|u\|^2_{\mathcal{W}}$$

where $\tilde{c}$ depends (only) on T^{-1}, $\|\mathcal{R}'\|_{\mathcal{L}(\mathcal{V},\mathcal{V}')}$, $\|\mathcal{R}\|_{\mathcal{L}(\mathcal{H})}$.

Now suppose to have ρ such that we can write

$$\Omega = \Omega_+(t) \cup \Omega_0(t) \cup \Omega_-(t) \qquad \text{for every } t \in [0,T],$$

where

$$\begin{aligned}\Omega_+(t) &= \{x \in \Omega \mid \rho(x,t) > 0\},\\ \Omega_-(t) &= \{x \in \Omega \mid \rho(x,t) < 0\},\\ \Omega_0(t) &= \{x \in \Omega \mid \rho(x,t) = 0\}.\end{aligned}$$

Denote by ρ_+ and ρ_- respectively the positive and negative parts of ρ. Notice that by i) of Definition 1 we can define the following spaces for every $t \in [0, T]$:

$$H_\diamond(t) := L^2\big(\Omega_\diamond(t); \rho_\diamond(t)\big), \qquad \diamond \in \{-, +\},$$
$$H_0(t) := L^2\big(\Omega_0(t)\big),$$
$$U_+(0) := W_0^{1,p}\big(\Omega_+(0) \cup \Omega_0(0)\big),$$
$$U_-(T) := W_0^{1,p}\big(\Omega_-(T) \cup \Omega_0(T)\big),$$

where $H_\diamond(t)$ is defined as the completion of $W_0^{1,p}(\Omega_\diamond(t))$ with respect to the norm

$$\int_\Omega u^2(x)\rho(x, t)\, dx\,.$$

We then suppose

$$U_+(0) \text{ dense in } H_+(0)\,, \qquad U_-(T) \text{ dense in } H_-(T)\,. \tag{16}$$

Remark 2 (Comments About Assumption (16)) Assumption (16) allows to consider the following situation: consider $\rho \geqslant 0$ independent of t and suppose $\rho > 0$ in Ω_+ and $\rho = 0$ in $\Omega_0 = \Omega \setminus \Omega_+$. The sets Ω_+ or Ω_0 may also be Cantor type sets, with positive measure, and, since in this case $\Omega_+ \cup \Omega_0 = \Omega$ assumption (16) is satisfied.

The analogous thing in general does not hold if $\rho > 0$ in Ω_+ and $\rho < 0$ in $\Omega_- = \Omega \setminus \Omega_+$, i.e. one could find Ω_+ a Cantor type set for which $\Omega = \Omega_+ \cup \Omega_-$, $\Omega_0 = \emptyset$, for which (16) does not hold.

Therefore to consider a forward-backward equation with $\rho \neq 0$ a.e. one needs to consider $\Omega_+(t)$ and $\Omega_-(t)$ more regular (for instance, $\Omega_+(t)$ and $\Omega_-(t)$ connected and with Lipschitz boundaries).

We refer to [25] for more details.

Now we consider the operator, defined in (5),

$$\mathcal{A} : \mathcal{V} \to \mathcal{V}'\,,$$
$$\mathcal{A}u := -\mathrm{div}\big(a(x, t, u, Du)\big) + b(x, t, u, Du) + c(x, t, u)\,.$$

with a, b, c satisfying (1), and, for given

$$f \in \mathcal{V}', \quad \varphi \in H_+(0), \quad \psi \in H_-(T)\,, \tag{17}$$

consider the two problems (the initial/final conditions make sense thanks to Proposition 1)

$$
(\mathrm{I})\begin{cases} \mathcal{P}_1 u = (\rho u)_t + \mathcal{A}u = f & \text{in } Q \\ u = 0 & \text{in } \partial\Omega \times (0,T) \\ u = \varphi & \text{in } \Omega_+(0) \\ u = \psi & \text{in } \Omega_-(T)\,, \end{cases}
\qquad
(\mathrm{II})\begin{cases} \mathcal{P}_2 u = \rho\, u_t + \mathcal{A}u = f & \text{in } Q \\ u = 0 & \text{in } \partial\Omega \times (0,T) \\ u = \varphi & \text{in } \Omega_+(0) \\ u = \psi & \text{in } \Omega_-(T)\,. \end{cases} \tag{18}
$$

Theorem 1 *Suppose (16) holds. Define the operator* $\mathcal{P}_1 : \mathcal{W} \to \mathcal{V}'$ *by* $\mathcal{P}_1 u = (\mathcal{R}u)' + \mathcal{A}u$ *where* $\mathcal{R}$ *is defined in* (8) *with* $\rho \in \mathcal{E}(C_1, C_2)$ *for* $C_1, C_2 \geqslant 0$, *and suppose* $\mathcal{A} : \mathcal{V} \to \mathcal{V}'$ *to be continuous. Suppose that there exist two constants* $\alpha, \beta > 0$ *such that (for* $p = 2$)

$$
\begin{aligned}
&\big\langle \mathcal{A}u - \mathcal{A}v + \tfrac{1}{2}(\mathcal{R}'u - \mathcal{R}'v), u - v \big\rangle_{\mathcal{V}' \times \mathcal{V}} \geqslant \alpha \|u - v\|^2_{\mathcal{V}}\,, \\
&\|\mathcal{A}u + \tfrac{1}{2}\mathcal{R}'u\|_{\mathcal{V}'} \leqslant \beta \|u\|_{\mathcal{V}}
\end{aligned} \tag{19}
$$

or for some $p \in (2, +\infty)$

$$
\begin{aligned}
&\langle \mathcal{A}u - \mathcal{A}v, u - v\rangle_{\mathcal{V}' \times \mathcal{V}} \geqslant \alpha \|u - v\|^p_{\mathcal{V}}\,, \qquad \|\mathcal{A}u\|_{\mathcal{V}'} \leqslant \beta \|u\|^{p-1}_{\mathcal{V}} \\
&\langle \mathcal{R}'u, u\rangle_{\mathcal{V}' \times \mathcal{V}} \geqslant 0
\end{aligned} \tag{20}
$$

for every $u, v \in \mathcal{V}$. *Then there is a constant* $c = c(\alpha, \beta, p)$ *(depending only on* α, β, p *and proportional to* $\alpha^{-\frac{1}{p}} + \alpha^{-\frac{p-1}{p}}$*) such that for every* $u \in \mathcal{W}$

$$
\begin{aligned}
\|u\|_{\mathcal{W}} \leqslant c \Big[&\|\mathcal{P}_1 u\|_{\mathcal{V}'} + \|\mathcal{P}_1 u\|^{1/(p-1)}_{\mathcal{V}'} + \|\rho_-^{1/2}(T) u(T)\|^{2(p-1)/p}_{L^2(\Omega)} + \\
&+ \|\rho_+^{1/2}(0) u(0)\|^{2(p-1)/p}_{L^2(\Omega)} + \|\rho_-^{1/2}(T) u(T)\|^{2/p}_{L^2(\Omega)} + \|\rho_+^{1/2}(0) u(0)\|^{2/p}_{L^2(\Omega)} \Big].
\end{aligned}
$$

Moreover for every $f \in \mathcal{V}'$, $\varphi \in H_+(0)$, $\psi \in H_-(T)$ *problem* (18)-(I) *has a unique solution.*

Theorem 2 *Suppose (16) holds. Define the operator* $\mathcal{P}_2 : \mathcal{W} \to \mathcal{V}'$ *by* $\mathcal{P}_2 u = \mathcal{R}u' + \mathcal{A}u$, *with* $\mathcal{R}$ *and* $\mathcal{A}$ *satisying the same assumptions as in Theorem 1.*

If we have that there are two constants $\alpha, \beta > 0$ *such that (for* $p = 2$)

$$
\begin{aligned}
&\big\langle \mathcal{A}u - \mathcal{A}v - \tfrac{1}{2}(\mathcal{R}'u - \mathcal{R}'v), u - v \big\rangle_{\mathcal{V}' \times \mathcal{V}} \geqslant \alpha \|u - v\|^2_{\mathcal{V}}\,, \\
&\|\mathcal{A}u - \tfrac{1}{2}\mathcal{R}'u\|_{\mathcal{V}'} \leqslant \beta \|u\|_{\mathcal{V}}
\end{aligned} \tag{21}
$$

or for some $p \in (2, +\infty)$

$$\begin{aligned} &\langle \mathcal{A}u - \mathcal{A}v, u - v\rangle_{\mathcal{V}' \times \mathcal{V}} \geqslant \alpha \|u - v\|^p_{\mathcal{V}}, \qquad \|\mathcal{A}u\|_{\mathcal{V}'} \leqslant \beta \|u\|^{p-1}_{\mathcal{V}} \\ &\langle \mathcal{R}'u, u\rangle_{\mathcal{V}' \times \mathcal{V}} \leqslant 0 \end{aligned} \tag{22}$$

for every $u, v \in \mathcal{V}$*, then there is a constant* $c = c(\alpha, \beta, p)$ *(depending only on* $\alpha, \beta,$ p *and proportional to* $\alpha^{-\frac{1}{p}} + \alpha^{-\frac{p-1}{p}}$*) such that for every* $u \in \mathcal{W}$

$$\|u\|_{\mathcal{W}} \leqslant c \Big[\|\mathcal{P}_2 u\|_{\mathcal{V}'} + \|\mathcal{P}_2 u\|^{1/(p-1)}_{\mathcal{V}'} + \|\rho_-^{1/2}(T)u(T)\|^{2(p-1)/p}_{L^2(\Omega)} +$$
$$+ \|\rho_+^{1/2}(0)u(0)\|^{2(p-1)/p}_{L^2(\Omega)} + \|\rho_-^{1/2}(T)u(T)\|^{2/p}_{L^2(\Omega)} + \|\rho_+^{1/2}(0)u(0)\|^{2/p}_{L^2(\Omega)} \Big].$$

Moreover for every $f \in \mathcal{V}'$, $\varphi \in H_+(0)$, $\psi \in H_-(T)$ *problem* (18)-(II) *has a unique solution.*

Notice that, under assumptions (1), coerciveness requirements about $\mathcal{A}$ are satisfied if $b \equiv 0$ and $c \geqslant 0$ or, if b is not identically zero, provided $c(x, t, u)u \geqslant m|u|^p$ with m great enough.

Two Concrete Examples of Equation with Discontinuous ρ Consider, for the sake of simplicity, $n = 1$, $\Omega = (0, T)$, so that $Q = (0, T) \times (0, T)$, $\mathcal{A}u = -u_{xx}$. Suppose $\rho = A$ in $\{x < t\}$, $\rho = B$ in $\{x > t\}$, $A, B \in \mathbf{R}$. In the following we consider two equations (initial/final data are not important) with $u = 0$ at $\partial\Omega \times (0, T)$.

Equation $(\rho u)' + \mathcal{A}u = f$ – A weak formulation is easily writable, indeed if $f \in L^2(Q)$ multiplying the equation by $\phi \in C^1_c(Q)$ one gets

$$-\iint_Q u\phi_t \rho \, dxdt + \iint_Q u_x \phi_x \, dxdt = \iint_Q f\phi \, dxdt \qquad \text{for every } \phi \in C^1_c(Q).$$

On the other hand, if one wants to write explicitly the equation distributing the time derivative one gets

$$\rho\, u_t - u_{xx} + \mathcal{R}'u = f,$$

where $\langle \mathcal{R}'u, u\rangle = (A - B)\int_0^T u^2(t, t)\, dt$ for $u \in \mathcal{V}$ (see also (11), in this case $\gamma(t) = t$).

In this case one can suppose $A - B \geqslant 0$, so that assumption $(20)_2$, and then also assumption $(19)_1$, is satisfied. Since the equation is linear we can also weaken the assumption: by (10) we get, if $A - B < 0$, that

$$\langle \mathcal{A}u + \frac{1}{2}\mathcal{R}'u, u\rangle \geqslant \left(1 + (A - B)\frac{C}{2}\right) \|u\|^2_{H^1_0(0,T)}.$$

Then $(19)_1$ may hold also for $A < B$, in this case for $A - B > -2/C$.

Equation $\rho u' + \mathcal{A}u = f$ – In this other case we need to suppose $A < B$ (see assumption $(22)_2$ about $\mathcal{R}'$), but, as above, one can assume some values for which $A > B$.

The equation is easy to be written, while in the weak formulation a strange term appears. Indeed in this case (suppose as before $f \in L^2(Q)$) one can see $\rho\, u'$ as $(\rho\, u)' - \mathcal{R}' u$ and get

$$
\begin{aligned}
- \iint_Q u\, \phi_t\, \rho\, dx dt + (B - A) \int_0^T u(x,x)\phi(x,x)\, dx + \iint_Q u_x \phi_x\, dx dt \\
= \iint_Q f \phi\, dx dt
\end{aligned}
$$

for every $\phi \in C_c^1(Q)$.

Notice as in both cases, taking the trace $u(x, x)$ in the diagonal of Q (the set where ρ is discontinuous) makes sense for a function $u \in \mathcal{W}$.

3 An Apparently Unnatural Inequality

For the result of this section we refer to [26].

If $\rho \equiv 1$ one has that

$$
\mathcal{W} \subset C^0([0,T]; L^2(\Omega))
$$

and the embedding is continuous, i.e. there is a constant $c > 0$, depending only on T, such that

$$
\max_{t \in [0,T]} \int_\Omega u^2(x,t)\, dx \leqslant c \left(\|u\|_{\mathcal{V}} + \|u'\|_{\mathcal{V}'} \right)^2 = c\, \|u\|^2_{\mathcal{W}} .
$$

If we consider a function $\rho \in \mathcal{E}$ one gets (by Proposition 1)

$$
\max_{t \in [0,T]} \left| \int_\Omega u^2(x,t) \rho(x,t)\, dx \right| \leqslant c \left(\|u\|_{\mathcal{V}} + \|(\mathcal{R}u)'\|_{\mathcal{V}'} \right)^2 = c\, \|u\|^2_{\mathcal{W}} .
$$

If ρ takes both positive and negative values the previous estimate is quite weak.

A natural question is then: may we get a stronger inequality? I.e., if $u \in \mathcal{W}$ may we hope to get the previous inequality with $|\rho|$ in the left hand side and $\mathcal{R}$ in the right hand side of the previous inequality?

This inequality does hold and one can prove that there is a constant $\tilde{c}$ such that

$$
\max_{t \in [0,T]} \int_\Omega u^2(x,t) |\rho|(x,t)\, dx \leqslant \tilde{c} \left(\|u\|_{\mathcal{V}} + \|(\mathcal{R}u)'\|_{\mathcal{V}'} \right)^2 = \tilde{c}\, \|u\|^2_{\mathcal{W}} . \qquad (23)
$$

First it was shown in some particular cases, that we cannot comment on here due to space limitations, but for which we refer to [23] and Proposition 2.8 in [18].

More in general the final word on the matter is the following (see [26]). Call Γ the interface where ρ changes sign: the previous inequality holds, provided Γ is regular enough, with a constant $\tilde{c}$ which is at least $2\,c$ (in fact greater if the interface Γ is not flat (see [26]). The increase on the constant is due to the technique of the proof: one consider the function u in the region where $\rho > 0$ and tries to "reflect" the function. This may let think that the constant $\tilde{c}$ is not sharp, since the reflection increases the value of the constant on the estimate (23) and the constant $\tilde{c}$ increases proportionally to the number of interfaces where ρ changes sign.

Anyway an example shows that if we have many interfaces the constant has to increase and if the number of interfaces is going to $+\infty$ we cannot find a constant $\tilde{c}$ for which an estimate like (23) holds. The idea (but see the example in the last section of [26]) is to consider $\rho = -1$ in $(-1, 0)$, $\rho = 1$ in $(0, 1)$, extend it by periodicity in $\mathbf{R}$, consider $\rho_h(x) := \rho(hx)$, $h \in \mathbf{N}$, and restrict each ρ_h to $\Omega = (-1, 1)$. Then $\mathcal{R}_h$ (the operators associated to ρ_h) satisfy $\lim_{h\to+\infty}(\mathcal{R}_h u)' = 0$ in $\mathcal{V}'$ and $|\rho_h| = 1$ for every $h \in \mathbf{N}$. If (23) would hold with $\tilde{c}$ independent of h we would get, taking the limit,

$$\max_{t\in[0,T]} \int_\Omega u^2(x,t)\,dx \leqslant \tilde{c}\,\|u\|_{\mathcal{V}},$$

which cannot hold.

Therefore inequality (23) does hold with a constant that does increase proportionally to the number of interfaces where ρ changes sign.

4 Time Regularity Results

For the first result of this section we refer to [20]. Theorem 4 may be obtained adapting the proof of Theorem 3.

In this section we consider

$$p = 2 \qquad \text{and} \qquad \mathcal{A} \textbf{ linear},$$

i.e.

$$\mathcal{A}u = -\text{div}\big(a\,Du\big) + b\,Du + c\,u,$$

with

$$a = a(x,t) \text{ matrix}, \quad b = b(x,t) \text{ vector}, \quad c = c(x,t) \text{ scalar}.$$

We suppose that a_{ij}, b_i, c belong to $L^\infty(\Omega \times (0, T)$ and admit a partial derivative with respect to t in $L^\infty(\Omega \times (0, T)$. We denote by $A(t)$ the "frozen" operator

$$H_0^1(\Omega) \ni u \mapsto -\mathrm{div}\big(a(\cdot, t)\, Du\big) + b(\cdot, t)\, Du + c(\cdot, t)u.$$

Then suppose that $[0, T] \ni t \mapsto \langle A(t)u, v\rangle_{H^{-1}(\Omega)\times H_0^1(\Omega)}$ is absolutely continuous and

$$\left| \frac{d}{dt} \langle A(t)u, v\rangle_{H^{-1}(\Omega)\times H_0^1(\Omega)} \right| \leqslant C_4 \, \|u\|_{H_0^1(\Omega)} \|v\|_{H_0^1(\Omega)} \tag{24}$$

for every $u, v \in H_0^1(\Omega)$, for a.e. $t \in [0, T]$ and for some non-negative constant C_4, i.e. a_{ij}, b_i, c are to be differentiable with respect to t with their temporal derivative bounded. In this way one can define $A'(t) \in \mathcal{L}(H_0^1(\Omega), H^{-1}(\Omega))$ and $\mathcal{A}' : \mathcal{V} \to \mathcal{V}'$, where $\mathcal{V}$ in this case is $L^2(0, T; H_0^1(\Omega))$ and where

$$\begin{aligned} A'(t)u = -\mathrm{div}\big(a'(\cdot, t)\, Du\big) + b'(\cdot, t)\, Du + c'(\cdot, t)u\,, \qquad & u \in H_0^1(\Omega), \\ \mathcal{A}u = -\mathrm{div}\big(a'\, Du\big) + b'\, Du + c'\, u\,, \qquad & u \in \mathcal{V}, \\ (a')_{ij} = \partial_t a_{ij}, \quad (b')_i = \partial_t b_i, \quad c' = \partial_t c, \qquad & 1 \leqslant i, j \leqslant n. \end{aligned}$$

As usual we will denote by $\mathcal{R}$ the operator associated to ρ and $R(t)$ the operator associated to $\rho(t)$ and, in accordance with the above notation, we will denote by ρ' the temporal derivate of ρ, if this should be differentiable, when we look at ρ as an abstract function $\rho : [0, T] \to L^\infty(\Omega)$.

Theorem 3 *Denote by u the solution of* (18)-(I). *Assume $\rho, \rho' \in \mathcal{E}(C_1, C_2)$, suppose that* (16) *and* (24) *to hold. Assume*

$$\alpha \|w\|^2_{H_0^1(\Omega)} \leqslant \langle [\rho'(t) + A(t)]w, w\rangle_{H^{-1}(\Omega), H_0^1(\Omega)} \leqslant \beta \|w\|^2_{H_0^1(\Omega)}$$

for every $t \in [0, T]$ and $w \in H_0^1(\Omega)$. Assume $f' \in \mathcal{V}'$, the existence of $u_0, u_T \in H_0^1(\Omega)$ such that

$$u_0 = \varphi \quad \text{in } \Omega_+(0) \qquad \text{and} \qquad u_T = \psi \quad \text{in } \Omega_-(T),$$

and the existence of $w_0, w_T \in L^2(\Omega)$ such that

$$f(0) - A(0)u_0 - \rho'(0)u_0 = \rho(0)w_0 \quad \text{and} \quad f(T) - A(T)u_T - \rho'(T)u_T = \rho(T)w_T.$$

Then $u' \in \mathcal{V}$, *i.e.* $u \in H^1(0,T;H_0^1(\Omega))$. *Moreover there is* $c > 0$ *such that*

$$\|u\|_{\mathcal{W}}+\|u'\|_{\mathcal{W}}+\sup_{t\in[0,T]}\|u(t)\|_{H_0^1(\Omega)}\leqslant$$

$$\leqslant c\Big[\|f\|_{\mathcal{V}'}+\|\rho_+^{1/2}(0)u_0\|_{L^2(\Omega)}+\|\rho_-^{1/2}(T)u_T\|_{L^2(\Omega)}+\|f'\|_{\mathcal{V}'}+$$

$$+\|\rho_+^{-1/2}(0)\big(f(0)-A(0)u_0-\rho'(0)u_0\big)\|_{L^2(\Omega_+(0))}+$$

$$+\|\rho_-^{-1/2}(T)\big(f(T)-A(T)u_T-\rho'(T)u_T\big)\|_{L^2(\Omega_-(T))}\Big].$$

In an analogous way one can derive also the following result.

Theorem 4 *Denote by* u *the solution of* (18)-(II). *Assume* $\rho, \rho' \in \mathcal{E}(C_1, C_2)$ *and suppose* (16) *and* (24) *to hold. Instead of* (19) *assume*

$$\alpha\|w\|^2_{H_0^1(\Omega)}\leqslant\langle A(t)w,w\rangle_{H^{-1}(\Omega),H_0^1(\Omega)}\leqslant\beta\|w\|^2_{H_0^1(\Omega)}$$

for every $t \in [0,T]$ *and* $w \in H_0^1(\Omega)$. *Assume* $f' \in \mathcal{V}'$ *and the existence of* $u_0, u_T \in H_0^1(\Omega)$ *such that*

$$u_0=\varphi \quad in\ \Omega_+(0) \qquad and \qquad u_T=\psi \quad in\ \Omega_-(T),$$

and the existence of $w_0, w_T \in L^2(\Omega)$ *such that*

$$f(0)-A(0)u_0=\rho(0)w_0 \qquad and \qquad f(T)-A(T)u_T=\rho(T)w_T.$$

Then $u' \in \mathcal{V}$. *Moreover there is* $c > 0$ *such that*

$$\|u\|_{\mathcal{W}}+\|u'\|_{\mathcal{W}}+\sup_{t\in[0,T]}\|u(t)\|_{H_0^1(\Omega)}\leqslant$$

$$\leqslant c\Big[\|f\|_{\mathcal{V}'}+\|\rho_+^{1/2}(0)u_0\|_{L^2(\Omega)}+\|\rho_-^{1/2}(T)u_T\|_{L^2(\Omega)}+\|f'\|_{\mathcal{V}'}+$$

$$+\|\rho_+^{-1/2}(0)\big(f(0)-Au_0\big)\|_{L^2(\Omega_+(0))}$$

$$+\|\rho_-^{-1/2}(T)\big(f(T)-Au_T\big)\|_{L^2(\Omega_-(T))}\Big].$$

Remark 3 Previous theorems says that, if a, b, c, ρ, f are differentiable with respect to t and their derivatives satisfy the stated assumptions, if moreover the initial/final data φ, ψ satisfy suitable compatibility conditions, then not only the solution $u \in \mathcal{W}$, but also $u' \in \mathcal{W}$, i.e. in particular both u and u' belongs to $\mathcal{V} = L^2(0,T;H_0^1(\Omega))$.

This means that $u \in H^1(0,T;H_0^1(\Omega))$ that continuously embeds in $C^0([0,T];H_0^1(\Omega))$.

Another result is the following, which is independent of the regularity of initial/final data φ and ψ. Consider the assumptions (in Theorem 3) $\rho, \rho' \in \mathcal{E}(C_1, C_2)$ and suppose (24) to hold. Assume moreover

$$\alpha \|w\|^2_{H^1_0(\Omega)} \leqslant \langle [\rho'(t) + A(t)]w, w\rangle_{H^{-1}(\Omega)\times H^1_0(\Omega)} \leqslant \beta \|w\|^2_{H^1_0(\Omega)}$$

for every $t \in [0, T]$ and $w \in H^1_0(\Omega)$ and assume $f' \in \mathcal{V}'$ (with no assumptions on initial/final data). Consider u the solution of (18)-(I). Then the function

$$v := t(T-t)u'$$

verifies ($\mathcal{P}_1$ defined in (18))

$$\begin{cases} \mathcal{P}_1 v = g := t(T-t)\big[f' - \mathcal{A}'u - (\mathcal{R}'u)'\big] + (T-2t)\mathcal{R}u' & \text{in } Q \\ v = 0 & \text{in } \partial\Omega \times (0, T) \\ v(0) = 0 & \text{in } \Omega_+(0) \\ v(T) = 0 & \text{in } \Omega_-(T). \end{cases}$$

Then, by Theorem 1,

$$v = t(T-t)u' \in \mathcal{V},$$

and there is $c > 0$ (depending on $\|t(T-t)\big[f' - \mathcal{A}'u - (\mathcal{R}'u)'\big] + (T-2t)\mathcal{R}u'\|_{\mathcal{V}'}$) such that

$$\|u\|_{\mathcal{V}} + \|t(T-t)u'\|_{\mathcal{V}} + \|(t(T-t)\mathcal{R}u')'\|_{\mathcal{V}'} \leqslant c\,.$$

Similarly, given u the solution of problem (18)-(II), the function

$$v := t(T-t)u'$$

verifies

$$\text{(II)} \begin{cases} \mathcal{P}_2 v = g := t(T-t)\big[f' - \mathcal{A}'u\big] + (T-2t)\mathcal{R}u' & \text{in } Q \\ v = 0 & \text{in } \partial\Omega \times (0, T) \\ v(0) = 0 & \text{in } \Omega_+(0) \\ v(T) = 0 & \text{in } \Omega_-(T) \end{cases}$$

and by Theorem 1 there is $c > 0$ (depending on $\|t(T-t)\big[f' - \mathcal{A}'u\big] + (T-2t)\mathcal{R}u'\|_{\mathcal{V}'}$) such that

$$\|u\|_{\mathcal{V}} + \|t(T-t)u'\|_{\mathcal{V}} + \|t(T-t)\mathcal{R}u''\|_{\mathcal{V}'} + \max_{t\in[0,T]} \|t(T-t)u'(t)\|_{H^1_0(\Omega)} \leqslant c\,.$$

Then, in particular, we have that for every $\delta > 0$, the solution u satisfies

$$u \in C^0([\delta, T-\delta]; H_0^1(\Omega)) \quad \text{and} \quad t(T-t)u \in C^0([0,T]; H_0^1(\Omega)).$$

5 Approximation via Elliptic Equations

For the result of this section we refer to [24].

In the following, we will take for granted the assumptions considered in Sect. 2 necessary for the existence of the solution of problems (18)-(I) and (18)-(II).

Problems (18) can be approximated via *elliptic* problems. This approach can be used also to give the existence result to problems (18): for instance, the solutions u_ε of problems (25) converge (as stated at the end of the present section) to some function u which solves (18)-(I) (and an analogous procedure could be done for problem (18)-(II)). This technique could be used to give existence result for problems (18) also in domains that are not cylindrical, similarly to what done in [21].

Consider f, φ, ψ as in (17), $\mathcal{A}$ as in (5) and the following *elliptic* problem

$$\begin{cases} -\varepsilon \dfrac{\partial}{\partial t}\left(\left|\dfrac{\partial u}{\partial t}\right|^{p-2} \dfrac{\partial u}{\partial t}\right) + \dfrac{\partial}{\partial t}\big(\rho(x,t)u\big) + \mathcal{A}u = f & \text{in } \Omega \times (0,T) \\ u = 0 & \text{in } \partial\Omega \times (0,T) \\ u = \varphi & \text{in } \Omega_+(0) \times \{0\} \\ \dfrac{\partial u}{\partial t}(x,0) = 0 & \text{in } \big(\Omega_0(0) \cup \Omega_-(0)\big) \times \{0\} \\ u = \psi & \text{in } \Omega_-(T) \\ \dfrac{\partial u}{\partial t}(x,T) = 0 & \text{in } \big(\Omega_0(T) \cup \Omega_+(T)\big) \times \{0\} \end{cases} \tag{25}$$

and the analogous one where in the equation one puts $\rho \frac{\partial u}{\partial t}$ instead of $\frac{\partial}{\partial t}(\rho\, u)$.

Notice that the additional boundary data are homogeneous Neumann data.

Consider

$$\mathcal{V}_* := \Big\{u \in \mathcal{V} \,\Big|\, u_t \in L^p(Q),\, u(\cdot,0) = 0 \text{ in } \Omega_+(0),\, u_t(\cdot,0) = 0 \text{ in } \Omega_0(0) \cup \Omega_-(0),$$
$$u(\cdot,T) = 0 \text{ in } \Omega_-(T),\, u_t(\cdot,T) = 0 \text{ in } \Omega_0(T) \cup \Omega_+(T)\Big\}.$$

Under assumptions of Theorem 1 problem (25) has a unique solution (see Theorem 3.6 in [24]). Denote by u_ε this solution and by v_ε the solutions (which exists under assumption of Theorem 2) of the same problems but where in the equation one puts $\rho \frac{\partial u}{\partial t}$ instead of $\frac{\partial}{\partial t}(\rho\, u)$. Then one has (see [24]):

$$\lim_{\varepsilon\to 0} \varepsilon\, \partial_t\big(|(u_\varepsilon)_t|^{p-2} (u_\varepsilon)_t \big) = \lim_{\varepsilon\to 0} \varepsilon\, \big(|(v_\varepsilon)_t|^{p-2} (v_\varepsilon)_t \big) = 0 \qquad \text{in } \mathcal{V}'_*,$$

$$\lim_{\varepsilon\to 0} u_\varepsilon = u, \quad \lim_{\varepsilon\to 0} v_\varepsilon = v \qquad \text{in } \mathcal{V},$$

$$\lim_{\varepsilon\to 0} (\rho\, u_\varepsilon)_t = (\rho\, u)_t, \quad \lim_{\varepsilon\to 0} \rho\, (v_\varepsilon)_t = \rho\, v_t \qquad \text{in } \mathcal{V}',$$

$$\|u_\varepsilon - u\|_{\mathcal{V}} = O(\varepsilon^{1/p}), \qquad \|v_\varepsilon - v\|_{\mathcal{V}} = O(\varepsilon^{1/p})$$

where u is the solution of (18)-(I) and v the solution of (18)-(II).

6 Local Boundedness and Harnack Inequality

In this section we confine to consider

$$\rho \geqslant 0 \qquad \text{and} \qquad p = 2.$$

As regard the result of local boundedness (for which we refer to [27]), this could be easily adapted to $p \geqslant 2$, while for the Harnack inequality (for which we refer to [16]) the case $p > 2$ is different and would require much more work.

The result of local boundedness for solutions if ρ takes both positive and negative values and $\rho \neq 0$ a.e. in Q is very delicate and for the moment has not been proved.

Returning to the case $\rho \geqslant 0$, once denoted by $\mathrm{Lip}(\Omega)$ the set of uniformly Lipschitz continuous functions in Ω and by $\mathrm{Lip}_c(\Omega)$ the subspace of functions belonging to $\mathrm{Lip}(\Omega)$ with compact support in Ω, we define the space

$$\mathcal{X}_c := \big\{\zeta \in \mathrm{Lip}(\Omega \times (0,T)) \,\big|\, \zeta(\cdot,t) \in \mathrm{Lip}_c(\Omega) \text{ for each } t \in (0,T),\ \zeta \geqslant 0,\ \zeta_t \geqslant 0\big\}.$$

One can prove that solutions of (3) or of (4) satisfies the following inequality for every function $\zeta \in \mathcal{X}_c$, for every $t_1, t_2 \in (0,T)$ and for every $k \in \mathbf{R}$

$$\begin{aligned}
&\int_\Omega (u-k)_+^2 \zeta^2 \rho(x,t_2)\, dx + \int_{t_1}^{t_2}\!\!\int_\Omega |D(u-k)_+|^2 \zeta^2 dxdt \leqslant \\
&\qquad \leqslant \int_\Omega (u-k)_+^2 \zeta^2 \rho(x,t_1) dx + \\
&\qquad\qquad + \gamma\Big[\int_{t_1}^{t_2}\!\!\int_\Omega (u-k)_+^2 |D\zeta|^2\, dxdt + \int_{t_1}^{t_2}\!\!\int_\Omega (u-k)_+^2 \zeta\zeta_t\, \rho\, dxdt + \\
&\qquad\qquad\qquad + \int_{t_1}^{t_2}\!\!\int_\Omega (u-k)_+^2 \zeta^2\, dxdt \\
&\qquad\qquad\qquad + k^2 \int_{t_1}^{t_2}\!\!\int_{\{u(t)>k\}} \big(\zeta^2 + |D\zeta|^2\big)\, dxdt \Big]
\end{aligned} \tag{26}$$

and the same inequality with $(u-k)_-$ at the place of $(u-k)_+$ and with

$$k^2 \int_{t_1}^{t_2}\int_{\{u(t)<k\}} (\zeta^2 + |D\zeta|^2)\, dxdt$$

in the place of the last term in inequality (26). We will refer to these inequalities as *energy inequalities.*

Functions satisfying these inequalities (or similar ones) are often said to belong to a De Giorgi class (of elliptic or parabolic type). For $\rho > 0$ the De Giorgi class is of parabolic type, for $\rho \equiv 0$ and u independent of t one can derive the analogous one in the elliptic case.

A first step was done studying the case

$$\rho > 0 \qquad \text{almost everywhere in } \Omega \times (0, T).$$

In this case a "weighted" local boundedness estimate and a "weighted" Harnack inequality are obtained. In particular, once defined

$$\mathsf{h}_*(x_o, t_o, r) := \frac{1}{r^2|B_r(x_o)|} \int_{t_o-r^2}^{t_o}\int_{B_r(x_o)} \rho(x,t)\, dxdt,$$

$$\mathsf{h}^*(x_o, t_o, r) := \frac{1}{r^2|B_r(x_o)|} \int_{t_o}^{t_o+r^2}\int_{B_r(x_o)} \rho(x,t)\, dxdt,$$

one has

$$\|u\|_{L^\infty(B_r(x_o)\times(t_o-\mathsf{h}_*(x_o,t_o,r)r^2,t_o))} \leqslant C\, \mathsf{M}\big(u;\, B_{\tilde r}(x_o) \times (t_o - \mathsf{h}_*(x_o, t_o, r)\,\tilde r^2, t_o)\big)$$

where $\tilde r > r > 0$, C depends (among other constants) on $(\tilde r - r)^{-1}$ and

$$\mathsf{M}(u;\, E) := \left(\frac{1}{\rho(E)} \iint_E u^2 \rho\, dx\, dt + \frac{1}{|E|} \iint_E u^2\, dx\, dt \right)^{1/2},$$

$$\rho(E) := \iint_E \rho\, dxdt,$$

and

$$u(x_o, t_o) \leqslant \eta \inf_{B_r(x_o)} u(x, t_o + r^2 \mathsf{h}^*(x_o, t_o, r))$$

for every u satisfying (26).

The word "weighted" means that to control u in a point (x_o, t_o) with the value of u in a ball $B_r(x_o)$ one has to "wait" a time proportional to the mean value of ρ in a suitable cylinder.

To address the case $\rho \geqslant 0$, in [27] a local boundedness result and in [16] a Harnack inequality are proved only for local solutions of (3) or (4) and not for functions satisfying (26). The reason why we have to confine only to solutions is explained below.

In both papers the result is proved under the assumption

$$Q = Q_+ \cup Q_0 \cup \Gamma,$$

where

$$Q_+ = \{\rho > 0\} \quad \text{and} \quad Q_0 = \{\rho = 0\}$$

are two non-empty disjoint connected and open subsets of Q, while Γ is the subset of codimension 1 (the set "where" ρ changes sign, from positive to null) defined by

$$\Gamma = \partial Q_+ \cap \partial Q_0$$

satisfying some assumptions that here we omit to report.

The proof is quite delicate and started by proving some inhomogeneous estimates. First one can modify ρ as follows, taking a function $\bar{\rho} > 0$ a.e. defined as

$$\bar{\rho} := \begin{cases} \rho & \text{in } Q_+, \\ \rho_0 & \text{in } Q_0, \end{cases} \qquad \bar{\rho} > 0 \text{ a.e.}$$

for a suitable ρ_0. Then one can prove inhomogeneous estimates, i.e. the two following estimates in two different regions (represented in Fig. 2)

$$\begin{aligned} \|u\|_{L^\infty(Q_r^+(x_o,t_o))} &\leqslant C\, \mathsf{M}(u; (Q_r^+)^\Delta(x_o,t_o)), \\ \|u\|_{L^\infty(Q_r^0(x_o,t_o))} &\leqslant C\, \mathsf{M}(u; (Q_r^0)^\Delta(x_o,t_o)), \end{aligned} \tag{27}$$

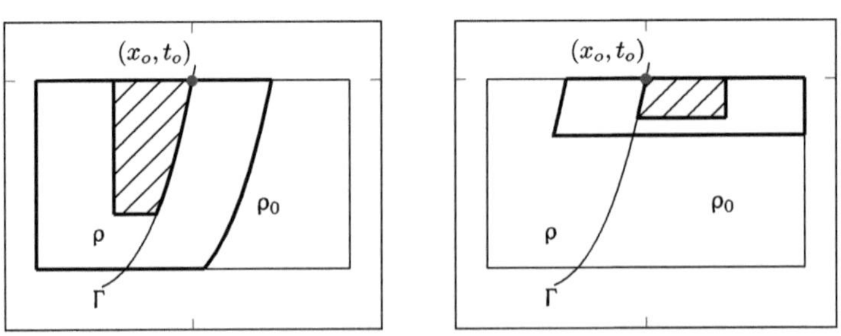

Fig. 2 An example of possible sets where estimates (27) hold

where

$$Q_r^+(x_o, t_o) = \bigcup_{t \in [t_o - r^2 \mathsf{h}_*^+(x_o, t_o, r), t_o]} B_r^+(x_o; t),$$

$$Q_r^0(x_o, t_o) = \bigcup_{t \in [t_o - r^2 \mathsf{h}_*^0(x_o, t_o, r), t_o]} B_r^0(x_o; t),$$

and, for $(x_o, t_o) \in \Gamma$,

$$\mathsf{h}_*^+(x_o, t_o, r) := \frac{\rho\big((B_r(x_o) \times (t_o - r^2, t_o)) \cap Q_+\big)}{\big|(B_r(x_o) \times (t_o - r^2, t_o)) \cap Q_+\big|},$$

$$\mathsf{h}_*^0(x_o, t_o, r) := \frac{\rho\big((B_r(x_o) \times (t_o - r^2, t_o)) \cap Q_0\big)}{\big|(B_r(x_o) \times (t_o - r^2, t_o)) \cap Q_0\big|}.$$

E^Δ is a suitable enlarged set containing E. In Fig. 2: the L^∞ norm in the dashed sets is estimated by the mean value of u^2 in the enlarged sets.

The second step consists of considering the estimates analogous to (27) with

$$\rho_\epsilon := \begin{cases} \rho & \text{in } Q_+, \\ \epsilon\, \rho_0 & \text{in } Q_0. \end{cases}$$

in the place of $\bar{\rho}$ and in a delicate passage to the limit in these estimates for $\epsilon \to 0^+$.

In the following picture, the L^∞ norm of a function u in the set in Fig. 3a, the union of the dashed set and the set

$$B_r^0(x_o; t_o) := (B_r(x_o) \times \{t_o\}) \cap Q_0,$$

is controlled by the mean value of u^2 in the set represented in Fig. 3b.

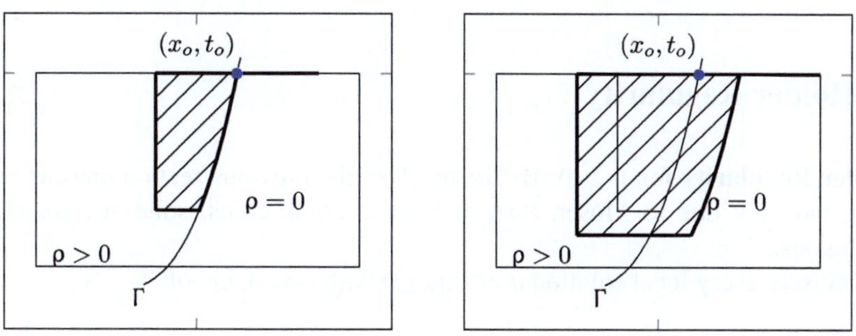

Fig. 3 The L^∞ norm in the set on the left hand side is estimated by the L^2 norm in the set on the right hand side

In particular in Q_0, taking the limit for $\epsilon \to 0^+$, one gets that for *almost every* t_o

$$\|u(\cdot, t_o)\|_{L^\infty(B^0_r(x_o))} \leqslant C\, \mathsf{M}_{t_o}(u;\, B^0_{\tilde{r}}(x_o))$$

where

$$\mathsf{M}_{t_o}(u;\, A) := \left(\frac{1}{\rho(t_o)(A)} \int_A u^2(x, t_o)\rho(x, \bar{t})\, dx + \frac{1}{|A|} \int_A u^2(x, t_o)\, dx \right)^{1/2} .$$

Coming to the Harnack inequality, this is much more delicate and difficult to be proved with respect to local boundedness, but the idea is similar (for this we refer to [16]).

The first step is to prove a general result for functions satisfying the energy inequalities (26) and the analogous one for $(u - k)_-$ with ρ_ϵ in the place of ρ, then take the limit as ϵ goes to 0. Also in this case the result holds only for solutions.

Here we state the inequality in the simplest case. Under suitable assumptions one can prove the following: consider a point $(x_o, t_o) \in \Gamma$, then there is a constant η such that for every non-negative local solution of (3) or (4) (with $\rho \geqslant 0$) and for almost every $t_o \in (0, T)$

$$\max\{ \sup_{B^+_r(x_o; t_o - r^2 \mathsf{h}^+_*(x_o, t_o, r))} u(x, t_o - r^2 \mathsf{h}^+_*(x_o, t_o, r)),\ \sup_{B^0_r(x_o; t_o)} u(x, t_o)\} \leqslant$$
$$\leqslant \eta \min\{ \inf_{B^+_r(x_o; t_o + r^2 \mathsf{h}^*_+(x_o, t_o, r))} u(x, t_o + r^2 \mathsf{h}^*_+(x_o, t_o, r)),\ \inf_{B^0_r(x_o; t_o)} u(x, t_o)\},$$

where $\mathsf{h}^+_*(x_o, t_o, r)$ and $B^0_r(x_o; t_o)$ are defined above and

$$\mathsf{h}^*_+(x_o, t_o, r) := \frac{\rho\big((B_r(x_o) \times (t_o, t_o + r^2)) \cap Q_+\big)}{\big|(B_r(x_o) \times (t_o, t_o + r^2)) \cap Q_+\big|},$$
$$B^+_r(x_o; t_o) := \big(B_r(x_o) \times \{t_o\}\big) \cap Q_+ .$$

7 Hölder Regularity

Hölder Regularity for $\rho \geqslant 0$ By the result of the previous section one can prove, using the trick due to Moser, that, at least in some cases, solutions are Hölder continuous.

Precisely every local solution u of Eqs. (3) with $c \equiv 0$, i.e. of

$$\rho(x, t)\frac{\partial u}{\partial t} - \operatorname{div}\big(a(x, t, u, Du)\big) + b(x, t, u, Du) = 0 \qquad \text{in } \Omega \times (0, T)\,,$$

is locally Hölder continuous in Q_+ both in space and time, is locally Hölder continuous in Γ both in space and time, while in Q_0 for almost every $t \in (0, T)$ the function $u(\cdot, t)$ is locally Hölder continuous in space.

Continuity also with respect to time is not to be be expected in Q_0, as an example in [16] shows.

Hölder Regularity for ρ > 0 Hölder continuity in the more general case, i.e. for both the Eqs. (3) and (4) (also with c not identically zero), for the moment is known only for

$$\rho > 0 \quad \text{a.e. in } \Omega \times (0, T).$$

For the precise result we refer to [17], where the case with $p = 2$ is considered.

The result could be not so expected since ρ can degenerate to 0 and overall because ρ, as well as the coefficients of $\mathcal{A}$, can be discontinuous.

8 *G*-Convergence

The result stated in this section can be found in [19] (clearly the assumptions for the existence of the solutions of Sect. 2 has to hold and $p \geqslant 2$).

Consider a sequence $(a_h)_{h\in\mathbf{N}}$ of Carathéodory functions

$$a_h : \Omega \times (0, T) \times \mathbf{R}^n \to \mathbf{R}^n\,, \qquad a_h = a_h(x, t, \xi)$$

satisfying

$$\text{(S.1)}\ a_h(x, t, 0) = 0\,,$$

$$\text{(S.2)}\ |a_h(x, t, \xi) - a_h(x, t, \eta)| \leqslant \Lambda\big(1 + |\xi|^p + |\eta|^p\big)^{(p-2)/p}|\xi - \eta|\,,$$

$$\text{(S.3)}\ (a_h(x, t, \xi) - a_h(x, t, \eta), \xi - \eta) \geqslant \lambda|\xi - \eta|^p\,.$$

Then one can prove the following: there is a function a satisfying (S.1), (S.3) and

$$\text{(S.2)}'\ |a(x, t, \xi) - a(x, t, \eta)| \leqslant \Lambda'(1 + |\xi|^p + |\eta|^p)^{\frac{p-2}{p-1}}|\xi - \eta|^{\frac{1}{p-1}}$$

($\Lambda' \geqslant \Lambda$ depends only on Λ and p), and a subsequence $(a_{h_j})_j$ such that for every

$$f \in L^{p'}(0, T; W^{-1,p'}(\Omega)), \quad \varphi, \psi \in L^2(\Omega)$$

and for every sequence $(\rho_h)_{h\in\mathbf{N}} \subset \mathcal{E}(C_1, C_2)$ with

$$\rho_h \to_h \rho \qquad \text{in } L^\infty(\Omega \times (0, T))\text{-weakly}*, \tag{28}$$

once denoted by $u_h = u_h(f, \varphi, \psi)$ the solution of

$$\begin{cases} \partial_t\big(\rho_h(x,t)v\big) - \mathrm{div}\big(a_h(x,t,Dv)\big) = f\,, & \text{in } \Omega \times (0,T)\,, \\ v = 0 & \text{in } \partial\Omega \times (0,T)\,, \\ v = \varphi & \text{in } \Omega^h_+(0) := \{x \in \Omega \mid \rho_h(\cdot,0) > 0\}\,, \\ v = \psi & \text{in } \Omega^h_-(T) := \{x \in \Omega \mid \rho_h(\cdot,T) < 0\}\,, \end{cases}$$

one has that

$$u_{h_j} \to_j u \qquad \text{in } L^p(0,T;L^2(\Omega)),$$

$$a_{h_j}(\cdot,\cdot,Du_{h_j}) \to_j a(\cdot,\cdot,Du) \qquad \textit{in } L^{p'}(0,T;(L^2(\Omega))^n)\textit{-weakly},$$

where u is the solution of the problem

$$\begin{cases} \partial_t\big(\rho(x,t)v\big) - \mathrm{div}\big(a(x,t,Dv)\big) = f\,, & \text{in } \Omega \times (0,T)\,, \\ v = 0 & \text{in } \partial\Omega \times (0,T)\,, \\ v = \varphi & \text{in } \Omega_+(0) := \{x \in \Omega \mid \rho(\cdot,0) > 0\}\,, \\ v = \psi & \text{in } \Omega_-(T) := \{x \in \Omega \mid \rho(\cdot,T) < 0\}\,. \end{cases}$$

Remark 4 Notice that once considered $(\rho_h)_{h\in\mathbf{N}} \subset \mathcal{E}(C_1, C_2)$ condition (28) is free, at least for a subsequence. Then, extracting a further subsequence if necessary, the result holds for every sequence $(\rho_h)_{h\in\mathbf{N}} \subset \mathcal{E}(C_1, C_2)$.

Remark 5 Notice that the "limit" function a is independent of the sequence $(\rho_h)_{h\in\mathbf{N}} \subset \mathcal{E}(C_1, C_2)$. Then, in particular, the "limit" function a is the same whether $\rho_h \equiv 1$ for every h, or $\rho_h \equiv 0$ for every h, or $(\rho_h)_{h\in\mathbf{N}}$ is any other sequence in $\mathcal{E}(C_1, C_2)$.

The proof needs many steps, but in particular two not trivial steps are the following:

- the first one is the compactness of the solutions $(u_h)_h$, that in the classical cases is free, since "where" ρ_h is zero one looses information about temporal derivative;
- the other one is the convergence of

$$U_h(t) := \int_\Omega u_h^2(x,t)\rho_h(x,t)\,dx \to_h U(t) := \int_\Omega u^2(x,t)\rho(x,t)\,dx$$
$$\text{in } C^0\big([0,T]\big)$$

and here the major difficulty if the fact that ρ_h may change sign, taking both positive and negative values.

Finally we want to observe that, with no many efforts, one could prove the analogous result for sequence of problems

$$\begin{cases} \rho_h(x,t)\dfrac{\partial v}{\partial t} - \operatorname{div}\big(a_h(x,t,Dv)\big) = f\,, & \text{in } \Omega \times (0,T)\,, \\ v = 0 & \text{in } \partial\Omega \times (0,T)\,, \\ v = \varphi & \text{in } \Omega_+^h(0) := \{x \in \Omega \mid \rho_h(\cdot,0) > 0\}\,, \\ v = \psi & \text{in } \Omega_-^h(T) := \{x \in \Omega \mid \rho_h(\cdot,T) < 0\}\,. \end{cases}$$

Remark 6 Due to assumptions for the existence of the solutions ($\mathcal{R}_h'$ needs to have a unique sign, being $\mathcal{R}_h$ the operator defined analogously to (8), see $(19)_2$, $(20)_2$, $(21)_2$, $(22)_2$), studying homogenisation can be done considering, e.g., $a_h(x,t,\xi) = a(h^\alpha x, h^\beta t, \xi)$, but only $\rho_h(x,t) = \rho(h^\gamma x, t)$, i.e. ρ_h non-oscillating in time, otherwise the operators $\mathcal{R}_h'$ would not have a unique sign and would not be uniformly bounded.

Acknowledgments The author wants to thank the referee for the numerous suggestions that have made this article more accessible and clear.

Funding Information This work has been partially supported by GNAMPA - INdAM 2025 Project *Metodi variazionali per problemi dipendenti da operatori frazionari isotropi e anisotropi.*

References

1. Baouendi, M.S., Grisvard, P.: Sur une équation d'evolution changeant de type. J. Funct. Anal. **2**, 352–367 (1968)
2. Beals, R.: On an equation of mixed type from electron scattering theory. J. Math. Anal. Appl. **58**, 32–45 (1977)
3. Beals, R.: An abstract treatment of some forward-backward problems of transport and scattering. J. Funct. Anal. **34**, 1–20 (1979)
4. Carroll, R.W., Showalter, R.E.: Singular and Degenerate Cauchy Problems. Academic Press, New York (1976)
5. Cercignani, C.: The Boltzmann Equation and Its Applications. Springer-Verlag, New York (1988)
6. Chanillo, S., Wheeden, R.L.: Weighted Poincaré and Sobolev inequalities and estimates for weighted Peano maximal functions. Am. J. Math. **107**(5), 1191–1226 (1985)
7. Glazatov, S.N.: Some nonclassical boundary value problems for linear equations of mixed type. Sibirsk. Mat. Zh. **44**, 44–52 (2003)
8. Goldstein, J.A., Mazumdar, T.: A heat equation in which the diffusion coefficient changes sign. J. Math. Anal. Appl. **103**, 533–564 (1984)
9. Gutiérrez, C.E., Wheeden, R.L.: Mean value and Harnak inequalities for degenerate parabolic equations. Colloq. Math. **60/61**, 157–194 (1990)
10. Gutiérrez, C.E., Wheeden, R.L.: Harnack's inequality for degenerate parabolic equations. Comm. Partial Differ. Equ. **16**(483), 745–770 (1991)
11. Karabash, I.M.: Abstract kinetic equations with positive collision operators. In: Spectral Theory in Inner Product Spaces and Applications, vol. 188. Operator Theory: Advances and Applications, pp. 175–195. Birkhäuser Verlag, Basel (2009)

12. Lions, J.L.: Quelques methodes de résolution des problème aux limites non linéaires. Gauthier-Villars, Paris (1969)
13. Pagani, C.D.: On forward-backward parabolic equations in bounded domains. Bollettino U.M.I. **13-B**(5), 336–354 (1976)
14. Pagani, C.D., Talenti, G.: On a forward-backward parabolic equation. Ann. Mat. Pura Appl. **90**, 1–57 (1971)
15. Pankov, A.A., Pankova, T.E.: Нелинейные еволюционные уравнения с необратимым операториым коеффициентом при производиой (Nonlinear evolution equations with non-invertible operator coefficient at the derivative). Doklady Akad. Nauk Ukr. **9**, 18–20 (1993)
16. Paronetto, F.: A Harnack inequality for solutions of elliptic-parabolic equations. Submitted
17. Paronetto, F.: Inner hölder continuity for parabolic de giorgi classes with degenerating coefficients depending on time. to appear in NoDEA
18. Paronetto, F.: Existence results for a class of evolution equations of mixed type. J. Funct. Anal. **212**(2), 324–356 (2004)
19. Paronetto, F.: G-convergence of mixed type evolution operators. J. Math. Pures Appl. (9) **93**, 361–407 (2010)
20. Paronetto, F.: A time regularity result for forward-backward parabolic equations. Boll. Unione Mat. Ital. (9) **4**(1), 69–77 (2011)
21. Paronetto, F.: An existence result for evolution equations in non-cylindrical domains. NoDEA Nonlin. Differ. Equ. Appl. **20**, 1723–1740 (2013)
22. Paronetto, F.: Local boundedness for forward-backward parabolic De Giorgi classes with coefficients depending on time. Nonlin. Anal. **158**, 168–198 (2017)
23. Paronetto, F.: A remark on forward-backward parabolic equations. Appl. Anal. **98**, 1042–1051 (2019)
24. Paronetto, F.: Elliptic approximation of forward-backward parabolic equations. Commun. Pure Appl. Anal. **19**, 1017–1036 (2020)
25. Paronetto, F.: Further existence results for elliptic-parabolic and forward-backward parabolic equations. Calc. Var. Partial Differ. Equ. **59**, 30 (2020). Paper No. 137
26. Paronetto, F.: An apparently unnatural estimate about forward-backward parabolic equations. Adv. Differ. Equ. **26**, 133–162 (2021)
27. Paronetto, F.: A natural local boundedness result for solutions of elliptic-parabolic equations. Anal. Appl. (Singap.) **24**(1), 67–96 (2026)
28. Pyatkov, S.G.: Solvability of a boundary value problem for a parabolic equation with changing time direction. Dokl. Akad. Nauk SSSR **285**, 1327–1329 (1985)
29. Showalter, R.E.: Degenerate parabolic initial-boundary value problems. J. Differ. Equ. **31**, 296–312 (1979)
30. Showalter, R.E.: Monotone Operators in Banach Space and Nonlinear Partial Differential Equations. American Mathematical Society, Providence (1997)
31. Ženišek, A.: Nonlinear Elliptic and Evolution Problems and Their Finite Element Approximations. Academic Press, London (1990)

A Survey on Anisotropic Integral Representation Results

Simone Verzellesi

Abstract In this chapter we review some recent results concerning integral representation properties of local functionals driven by Lipschitz continuous anisotropies.

1 Introduction

A *local functional* is a functional F of two variables: a function u belonging to a chosen functional space, a set A belonging to a chosen family of sets. The adjective *local* refers to the fact that $F(u, A)$ depends only on the values that u takes on A. An important class of local functionals is that of *integral functionals*, e.g. functionals of the form

$$(u, A) \mapsto \int_A f(x, u, Du)\, dx,$$

where, say, A is open and $u \in W^{1,p}(A)$. Integral functionals constitute the cornerstone of the modern calculus of variations, and describing their importance clearly goes beyond our scope. We refer to [16, 26, 31] for complete introductions to this topic. In the study of optimization problems for local functionals, it is now an established practice to rely on the tool of Γ-*convergence*, as introduced by E. De Giorgi and T. Franzoni [19, 20]. For an exhaustive introduction to this topic, we refer to the monographs [8, 9, 18]. Since the late 1970s, G. Buttazzo and G. Dal Maso have investigated Γ-convergence in the framework of Lebesgue spaces, Sobolev spaces and BV spaces [11, 13, 17]. A key tool in the study of Γ-convergence properties in these frameworks consists in the so-called *integral representation*. By integral representation we mean finding conditions under which a local functional

S. Verzellesi (✉)
Dipartimento di Matematica "Tullio Levi-Civita", Università degli Studi di Padova, Padova, Italy
e-mail: simone.verzellesi@unipd.it

J. Duran i Lamiel, A. Maione (eds.), *New Frontiers in Homogenization and Fractional Calculus*, Trends in Mathematics,
https://doi.org/10.1007/978-3-032-20098-3_8

$F(u, A)$ can be expressed e.g. as

$$F(u, A) = \int_A f(x, u, Du)\, dx$$

for a suitable *Lagrangian* f. In the Euclidean setting this problem is well understood, and we refer the reader to [3, 10–13] for an overview of the subject. In this chapter, we review some recent results concerning local functionals driven by Lipschitz continuous anisotropies, tailored to study anisotropic integral functionals of the form

$$F(u, A) = \int_A f(x, u, Xu)\, dx, \qquad A \subseteq \Omega,\ u \in W^{1,p}_X(\Omega), \tag{1}$$

where $W^{1,p}_X(\Omega)$ is an appropriate anisotropic Sobolev space. An *anisotropy* is a family $X = (X_1, \dots, X_m)$ of Lipschitz continuous vector fields inducing the anisotropic gradient

$$Xu = (X_1 u, \dots, X_m u) = X_1 u X_1 + \dots + X_m u X_m.$$

The term *anisotropy* is motivated by the fact that X may vary point by point. The investigation of anisotropic variational functionals as in (1) is originally motivated by L. Hörmander's seminal work on *hypoelliptic operators* [34]. The latter constitutes a milestone in the study of differential operators with underlying *sub-Riemannian* type structures. Important evidences are the works of G.B. Folland [24] and of L. Rothschild and E.M. Stein [43] in the context of *stratified* and *nilpotent Lie groups*. We refer to [2, 7, 33, 35] for further accounts. The archetypal example consists in the smooth anisotropy $X = (X_1, X_2)$ on $\mathbf{R}^3$, where

$$X_1 = \frac{\partial}{\partial x_1} + x_2 \frac{\partial}{\partial x_3} \qquad \text{and} \qquad X_2 = \frac{\partial}{\partial x_2} - x_1 \frac{\partial}{\partial x_3}. \tag{2}$$

The latter generate, via Lie-brackets, the Lie algebra of the so-called *first Heisenberg group*, and induce on it a *sub-Riemannian structure*. The relevant hypoelliptic operator associated with X is the so-called *sub-Laplacian*

$$(X_1)^2 u + (X_2)^2 u, \tag{3}$$

whose associated *Dirichlet energy* reads as

$$\int_\Omega \left(|X_1 u|^2 + |X_2 u|^2 \right) \mathrm{d}x. \tag{4}$$

As (3) is not elliptic, (4) is not coercive. Moreover, the Euclidean Sobolev spaces $W^{1,2}(\Omega)$ are not the correct finiteness domains for functionals as in (3). These issues are clear obstructions to the use of classical techniques to study minimization properties of (4), preventing for instance (cf. [45]) the L^2-lower semicontinuity of its Euclidean extension to $L^2(\Omega)$, namely

$$\begin{cases} \int_\Omega \left(|X_1 u|^2 + |X_2 u|^2\right) \mathrm{d}x & \text{if } u \in W^{1,2}(\Omega), \\ \infty & \text{if } u \in L^2(\Omega) \setminus W^{1,2}(\Omega). \end{cases}$$

To overcome these obstacles, building on the foundational work of Folland and Stein [25], the correct functional framework has been developed by B. Franchi et al. [27, 28] and by N. Garofalo and D.M. Nhieu [29, 30]. Precisely, when $1 \leq p < \infty$, the study of anisotropic functionals in the generality of (1) can be carried out via the introduction of anisotropic Sobolev and BV spaces, say $W^{1,p}_X(\Omega)$ and $BV_X(\Omega)$, emerging as the natural domains of functionals as in (1). When $p = \infty$, the reader is referred to [14, 42, 46, 47] for some anisotropic L^∞-variational problems.

The study of anisotropic integral representation results started in [37, 38], where the authors investigated integral representation and Γ-convergence properties of functionals $F(u, A)$ modeled by suitable anisotropies $X = (X_1, \dots, X_m)$, assuming that

$$X_1(x), \dots, X_m(x) \text{ are linearly independent for a.e.} x \in \Omega \tag{LIC}$$

and

$$F(u + c, A) = F(u, A) \text{ for any function } u, \text{ any set } A \text{ and any constant} c. \tag{TI}$$

According to [37], we call a functional *translation-invariant* whether it satisfies (TI), while we say that X satisfies the *linear independence condition* if it satisfies (LIC). Local functionals satisfying (TI) model integral functionals of the form

$$F(u, A) = \int_A f(x, Du)\, dx, \tag{5}$$

i.e. whose Lagrangian is independent of u. Although not fully general, the study of functionals satisfying (TI) is nevertheless highly relevant in the literature. Therefore, we will limit ourselves mainly to this class. On the other hand, (LIC) already embraces many relevant families of vector fields studied in literature (cf. [38] for some instances). In [37] the authors found conditions under which F can be represented as

$$F(u, A) = \int_A f(x, Xu)\, dx \qquad \text{for any } A \subseteq \Omega \text{ open, any } u \in W^{1,p}_{X,loc}(A) \cap L^p(\Omega), \tag{6}$$

and for a Lagrangian $f : \Omega \times \mathbf{R}^m \to [0, \infty)$. Moreover, they applied this characterization to prove a Γ-compactness theorem for integral functionals of the form (6), when $1 < p < \infty$. We refer to [36] for similar results under stronger conditions. These results were later extended in [22, 23] beyond (TI), to model integral functionals of the form

$$F(u, A) = \int_A f(x, u, Xu)\, dx \qquad \text{for any } A \subseteq \Omega \text{ open, any } u \in W^{1,p}_{X,loc}(A) \cap L^p(\Omega),$$

and generalizing their Euclidean counterparts [11, 13]. Although the above-mentioned contributions relies on the linear independence condition (LIC), the latter has been definitely removed in [45], allowing for anisotropic integral representation results in the greatest generality.

In this chapter we review the approach to the above integral representations, focusing on the case in which (TI) holds. First, we recall the strategy developed by Buttazzo and Dal Maso to prove the Euclidean result. Their approach can be summarized into the following three steps.

1. Prove (5) when u is a piecewise affine function. This step require the fact that F is a measure in its second variable.
2. Recall that piecewise affine functions are dense in Sobolev spaces.
3. Exploit the above steps, together with lower semicontinuity properties of F, to extend (5) to all Sobolev functions.

This scheme, which applies beyond (TI), relies heavily on the second step. However, as noticed in [37], it is not generally the case that anisotropic piecewise affine functions approximate anisotropic Sobolev functions. To this aim, the authors of [37] developed the following alternative strategy.

1. Apply the Euclidean integral representation result to the functional, obtaining an integral representation with respect to a Euclidean Lagrangian f_e of the form

$$F(u, A) = \int_A f_e(x, u, Du)\, dx \quad \text{for any } A \subseteq \Omega \text{ open and } u \in W^{1,p}_{loc}(A) \cap L^p(\Omega).$$

2. Find conditions on f_e guaranteeing the existence of a Lagrangian f such that

$$\int_A f_e(x, u, Du)\, dx = \int_A f(x, u, Xu)\, dx \quad \text{for any } A \subseteq \Omega \text{ open and } u \in C^\infty(A). \tag{7}$$

3. Combine the previous steps with density properties of smooth functions.

While the third step relies on well-established anisotropic Meyers-Serrin-type approximation results [27, 29], the second one constitutes the key of this approach. Indeed, its achievement led the authors of [37] to assume (LIC), and the author of [45] to remove this assumption. We point out that, although (LIC) is in the end not necessary to establish (7), it nevertheless plays a role in certain secondary issues, including, for example, the uniqueness of representation.

Our exposition is organized as follows. In Sect. 2 we recall the background about the anisotropic functional framework (Sect. 2.2), introducing the relevant Sobolev spaces (Sect. 2.3) and discussing density properties of piecewise affine and smooth functions (Sect. 2.4). In Sect. 3, we introduce the basic terminology of local functionals. In Sect. 4 we prove the integral representation results in the (TI) setting. Precisely, first we recall the Euclidean approach (Sect. 4.1), then we explain how to establish (7) (Sect. 4.2), and finally we complete the proof of the anisotropic integral representation result (Sect. 4.3). In Sect. 5 we explain when and how (LIC) is actually relevant. Finally, in Sect. 6 we guide the reader to existing literature, including cases beyond (TI) and applications to Γ-convergence.

2 Anisotropic Functional Frameworks

2.1 Main Notations

We let $\mathbf{N} = \mathbf{N} \setminus \{0\}$. We fix $m, n \in \mathbf{N}$. For $\alpha, \beta \in \mathbf{N}$, we denote by $M(\alpha, \beta)$ the set of matrices with α rows and β columns. We denote by $\langle \cdot, \cdot \rangle$ the Euclidean scalar product, and by $| \cdot |$ its associated norm. If $L : \mathbf{R}^\alpha \to \mathbf{R}^\beta$ is a linear map, we denote by $\ker(L) \subseteq \mathbf{R}^\alpha$ and $\mathrm{Im}(L) \subseteq \mathbf{R}^\beta$ its kernel and its range. Fixed an open and bounded set $\Omega \subseteq \mathbf{R}^n$, $\mathcal{A}$ is the class of open subsets of Ω, and $\mathcal{B}$ is the class of Borel subsets of Ω. We let $\mathcal{A}_0$ be the subfamily of $\mathcal{A}$ of open subsets A of Ω such that $A \Subset \Omega$, and by $\mathcal{B}_0$ the subfamily of $\mathcal{B}$ of Borel subsets B of Ω such that $B \Subset \Omega$. Finally, we denote by Du the (distributional) Euclidean gradient of u. Gradients may be either columns or rows.

2.2 Anisotropies

In the following, Ω is an open and bounded subset of $\mathbf{R}^n$. Given a family $X = (X_1, \ldots, X_m)$ of Lipschitz continuous vector fields on Ω, such that

$$X_j = \sum_{i=1}^{n} c_{j,i} \frac{\partial}{\partial x_i}, \qquad \text{with } c_{j,i} \in \mathrm{Lip}(\Omega)$$

for any $j = 1, \ldots, m$ and any $i = 1, \ldots, n$, we denote by $C(x) := [c_{j,i}(x)]_{\substack{i=1,\ldots,n \\ j=1,\ldots,m}}$ the $m \times n$ *coefficient matrix* associated with X. We may refer to X as *anisotropy* or X-*gradient*. This notation is motivated by the following definition.

Definition 1 Let u and V be a one-dimensional and an m-dimensional distribution.

1. The X-gradient of u is the m-dimensional distribution defined by

$$Xu(\varphi) = -u\left(\operatorname{div}(C^T\varphi)\right) \qquad \text{for any } \varphi \in \mathbf{C}_c^\infty(\Omega; \mathbf{R}^m).$$

2. The X-divergence of V is the one-dimensional distribution defined by

$$\operatorname{div}_X(V)(\varphi) = -V(CD\varphi) \qquad \text{for any } \varphi \in \mathbf{C}_c^\infty(\Omega).$$

Below are some examples of relevant anisotropies.

Example 1 (The Euclidean Space) When $m = n$ and $X_j = \frac{\partial}{\partial x_j}$, we are in the Euclidean isotropic setting, and we feel there is no need to add more.

Example 2 (Riemannian Structures) When $m = n$ and X is a global frame of smooth vector fields, Ω can be endowed with a Riemannian metric for which X is an orthonormal frame. Then X, div_X and d are the Riemannian gradient, divergence and distance respectively.

Example 3 (Carnot Groups) Let $n = 3$, $m = 2$, $X_1 = \frac{\partial}{\partial x_1} + x_2\frac{\partial}{\partial x_3}$ and $X_2 = \frac{\partial}{\partial x_2} - x_1\frac{\partial}{\partial x_3}$. In this case, X is a basis of left-invariant vector fields of the *first Heisenberg group*, a particular Lie group which constitute the first non-trivial instance in the wider class of *Carnot groups*.

Example 4 (Bracket-Generating Anisotropies) In Example 3,

$$[X_1, X_2] = -2\frac{\partial}{\partial x_3},$$

whence the Lie algebra generated by X has full rank at every point. Vector fields satisfying this property are called *bracket-generating*, and are the building blocks of *sub-Riemannian geometry*. Not every bracket-generating anisotropy gives rise to a Carnot group. The simplest instance is the *Grushin plane*, i.e. $n = m = 2$, $X_1 = \frac{\partial}{\partial x_1}$ and $X_2 = x_1\frac{\partial}{\partial x_2}$. Indeed,

$$[X_1, X_2] = \frac{\partial}{\partial x_2}.$$

Example 5 (Carnot-Carathéodory Spaces) A tool associated with anisotropies is the so-called *Carnot-Carathéodory distance* induced by X on Ω (cf. e.g. [33]). Precisely, let $x, y \in \Omega$, and let $\gamma : [0, 1] \to \Omega$ be any absolutely continuous curve joining x and y. γ is *horizontal if*

$$\dot{\gamma}(t) = \sum_{j=1}^{m} a_j(t) X_j(\gamma(t)) \qquad \text{for a.e. } t \in [0, 1]. \tag{8}$$

Then the Carnot-Carathéodory distance is defined by

$$d(x, y) = \inf\left\{ \int_0^1 |(a_1, \ldots, a_m)|\, dt \;:\; \gamma(0) = x, \gamma(1) = y, \text{ (8) holds} \right\}.$$

Depending on the anisotropy, d may not be always finite. If it is the case, it is a length distance, and (Ω, d) is called *Carnot-Carathéodory space* [40]. By *Chow-Rashevskii connectivity theorem* (cf. [15] and [35, Chapter 3]), bracket-generating anisotropies give rise to finite Carnot-Carathéodory distances. Nevertheless, this condition is not necessary. As an instance (cf. [45]), consider the family $X = (X_1, X_2)$ of vector fields defined on $\Omega = (-1, 1)^2 \subseteq \mathbf{R}^2$ by

$$X_1(x) = \frac{\partial}{\partial x_1}, \quad X_2(x) = \begin{cases} 0 & \text{if } x_1 \in (-1, 0) \\ x_1 \frac{\partial}{\partial x_2} & \text{if } x_1 \in [0, 1) \end{cases} \quad \text{for any } x = (x_1, x_2) \in \Omega.$$

Ω is easily a Carnot-Carathéodory space, while X is not bracket-generating.

2.3 Sobolev Spaces

The anisotropic Sobolev space $W^{1,p}_X(\Omega)$ is defined by

$$W^{1,p}_X(\Omega) = \{u \in L^p(\Omega) \;:\; Xu \in L^p(\Omega; \mathbf{R}^m)\}.$$

The space $W^{1,p}_{X,loc}(\Omega)$ is defined as usual. It is well-known (cf. [25]) that $W^{1,p}_X(\Omega)$, endowed with the norm

$$\|u\|_{W^{1,p}_X(\Omega)} := \|u\|_{L^p(\Omega)} + \|Xu\|_{L^p(\Omega;\mathbf{R}^m)},$$

is a Banach space for any $1 \le p \le \infty$, reflexive when $1 < p < \infty$. The Lipschitz continuity assumption on the family X ensures that $W^{1,p}(\Omega)$ embeds continuously into $W^{1,p}_X(\Omega)$ (cf. [23, 37]), where $W^{1,p}(\Omega)$ denotes the Euclidean Sobolev space. Precisely, for any $u \in W^{1,p}(\Omega)$, the X-gradient admits the Euclidean representation

$$Xu(x) = C(x) Du(x) \qquad \text{for a.e. } x \in \Omega. \tag{9}$$

However, in general, $W^{1,p}(\Omega) \subsetneq W_X^{1,p}(\Omega)$, as the next example shows (cf. [45]).

Example 6 Let $X = (X_1, X_2)$ be as in Example 3. Let $\Omega \subseteq \mathbf{R}^3$ be the cylinder defined by $\Omega = \left\{(x_1, x_2, x_3) \in \mathbf{R}^3 : x_1^2 + x_2^2 < 1,\ |x_3| < 1\right\}$. Let $F \in C_c^1(\mathbf{R})$ be such that supp$F \subseteq (-1, 1)$ and $F(s) = s$ on $\left(-\frac{1}{2}, \frac{1}{2}\right)$. Set $k = \sup_{s\in(-1,1)} \left|F'(s)\right|$. Define $u : \Omega \to \mathbf{R}$ by

$$u(x_1, x_2, x_3) := \begin{cases} F\left(\dfrac{x_3}{x_1^2 + x_2^2}\right) & \text{if } (x_1, x_2) \neq (0, 0) \\ 0 & \text{if } (x_1, x_2) = (0, 0) \end{cases}.$$

Then $u \in W_X^{1,2}(\Omega) \setminus W^{1,2}(\Omega)$ (cf. [45]).

2.4 Approximation Properties

For what concerns density of smooth functions, the classical result of Meyers and Serrin is still valid in the anisotropic setting and it was proved, independently, in [27] and [30].

Theorem 1 *For any* $1 \leq p < \infty$, $W_X^{1,p}(\Omega) = \overline{\mathbf{C}^\infty(\Omega) \cap W_X^{1,p}(\Omega)}^{\|\cdot\|_{W_X^{1,p}(\Omega)}}$.

Under additional assumptions on X, Poincaré inequalities and Rellich-Kondrachov-type theorems may be deduced. We refer the reader to [38, Proposition 2.16] and [28, Theorem 3.4], respectively. Instead, when we come to density of affine functions, the Euclidean and the anisotropic framework differ. In the Euclidean setting, it is well-known that piecewise affine functions approximate Sobolev functions. We recall that $u : \Omega \to \mathbf{R}$ is *piecewise affine* if $u \in C^0(\overline{\Omega})$ and if there exists a finite family $\{\Omega_1, \ldots, \Omega_s\} \subseteq \mathcal{A}$ and a negligible set N such that

$$\Omega = \bigcup_{i=1}^{s} \Omega_i \cup N \qquad \text{and} \qquad u|_{\Omega_i} \text{ is affine for any } i = 1, \ldots, s.$$

The following density result holds.

Proposition 1 *Let* $1 \leq p < \infty$. *Let* $u \in W^{1,p}(\Omega)$ *and let* $A' \Subset \Omega$. *Then there exists a sequence* $(u_h)_h$ *of piecewise affine functions converging to* u *strongly in* $W^{1,p}(A')$.

Proposition 1 is a corollary of the following result (cf. [21, Proposition 2.8]).

Proposition 2 *Let* $1 \leq p < \infty$. *Let* $\tilde{\Omega}$ *be an open and bounded subset of* $\mathbf{R}^n$ *with Lipschitz boundary. Let* $u \in W_0^{1,p}(\tilde{\Omega})$. *There exists a sequence* $(u_h)_h$ *of piecewise affine functions on* $\tilde{\Omega}$, *converging to* u *strongly in* $W^{1,p}(\tilde{\Omega})$, *such that* $u_h|_{\partial\tilde{\Omega}} = 0$ $h \in \mathbf{N}$.

Recall that u is affine if and only if Du is constant. If we call a smooth function u *X-affine* as soon as Xu is constant, it is natural to wonder whether the above properties hold for anisotropic Sobolev functions. As shown in [37], the answer is negative. The notion of piecewise X-affine function is given *verbatim* as above. Let Ω be a bounded open subset of $\mathbf{R}^3$. Let X_1, X_2 be as in Example 3. Let $u(x_1, x_2, x_3) = x_3$. Clearly $u \in C^\infty(\overline{\Omega})$, and so in particular $u \in W_X^{1,p}(\Omega)$ for any $p \geq 1$. An easy computation reveals that a function v is X-affine if and only if

$$v(x_1, x_2, x_3) = ax_1 + bx_2 + c$$

for some $a, b, c \in \mathbf{R}$. Therefore, since X-affine functions do not depend on x_3, u cannot be pointwise approximated by piecewise X-affine functions.

3 Local Functionals and Carathéodory Functions

3.1 Local Functionals

In this section we collect some definitions about increasing set functions and local functionals (cf. [18]). The latter, heuristically, behave like variational functionals in the first entry and like measures in the second one.

Definition 2 We say that $\alpha : \mathcal{A} \to [0, \infty]$ is

1. *increasing* if $\alpha(A) \leq \alpha(B)$ for any $A, B \in \mathcal{A}$ such that $A \subseteq B$;
2. *inner regular* if it is increasing and $\alpha(A) = \sup\{\alpha(A') : A' \Subset A\}$ for any $A \in \mathcal{A}$;
3. *subadditive* if it is increasing and $\alpha(A) \leq \alpha(B) + \alpha(C)$ for any $A, B, C \in \mathcal{A}$ such that $A \subseteq B \cup C$;
4. *superadditive* if it is increasing and $\alpha(C) \geq \alpha(A) + \alpha(B)$ for any $A, B, C \in \mathcal{A}$ with $A \cap B = \emptyset$ and $A \cup B \subseteq C$;
5. a *measure* if it is increasing and it is the restriction to $\mathcal{A}$ of a Borel measure.

The next result (cf. [18, Theorem 14.23]) gives conditions to be a measure.

Theorem 2 *Let $\alpha : \mathcal{A} \to [0, \infty]$ be an increasing function such that $\alpha(\emptyset) = 0$. Define the function $\alpha^* : \mathcal{B} \to [0, \infty]$ by*

$$\alpha^*(B) := \inf\{\alpha(A) : A \in \mathcal{A},\ B \subseteq A\}.$$

Then the following conditions are equivalent:

1. α is a measure;
2. α is subadditive, superadditive and inner regular;
3. α^* is a Borel measure coinciding with α on $\mathcal{A}$.

The next result is an immediate corollary of Theorem 2.

Corollary 1 *Let $\alpha : \mathcal{A} \to [0, \infty]$ be a measure. The following are equivalent:*

1. *For any $A' \in \mathcal{A}_0$, $\alpha(A') < \infty$;*
2. *α is a Radon measure.*

The following definition adapts some standard notation to the anisotropic setting.

Definition 3 If $F : L^p(\Omega) \times \mathcal{A} \to [0, \infty]$, we say that F is:

1. a *measure* if $F(u, \cdot)$ is a measure for any $u \in L^p(\Omega)$;
2. *local* if, for any $A \in \mathcal{A}$ and $u, v \in L^p(\Omega)$, then

$$u|_A = v|_A \implies F(u, A) = F(v, A);$$

3. *convex* if $F(\cdot, A)$ restricted to $W^{1,p}_{X,loc}(A) \cap L^p(\Omega)$ is convex for any $A \in \mathcal{A}$;
4. *L^p-lower semicontinuous* (respectively *$W^{1,p}_X$-lower semicontinuous*) if $F(\cdot, A)$ is strongly L^p-lower semicontinuous (respectively $W^{1,p}_X$-lower semicontinuous) for any $A \in \mathcal{A}$;
5. *weakly-$\star$ sequentially lower semicontinuous* if $F(\cdot, A)$ restricted to $W^{1,\infty}(\Omega)$ is weakly-$\star$ sequentially lower-semicontinuous for any $A \in \mathcal{A}$.

We point out that many of the above notions are not relevant in the (TI) setting.

3.2 *Carathéodory Functions*

In order to ensure that an integral functional of the form

$$F(u, A) = \int_A f(x, u, Xu)\, dx$$

is well-defined, we impose some conditions on the Lagrangian f in such a way that

$$x \mapsto f(x, u(x), Xu(x))$$

is integrable on A for any $u \in W^{1,1}_{X,loc}(A) \cap L^p(\Omega)$. To this aim, it is customary to work in the class of *Carathéodory functions* (cf. [16]).

Definition 4 (Carathéodory Functions) Let $f : \Omega \times \mathbf{R} \times \mathbf{R}^m \to [0, \infty]$ be a function. We say that f is a *Carathéodory function* if:

(i) $f(\cdot, u, \eta)$ is measurable for any $u \in \mathbf{R}$ and any $\eta \in \mathbf{R}^m$;
(ii) $f(x, \cdot, \cdot)$ is continuous for a.e. $x \in \Omega$.

Carathéodory functions constitute the right class of Lagrangians, as the next proposition shows (cf. e.g. [21])

Proposition 3 *Let $f : \Omega \times \mathbf{R} \times \mathbf{R}^m \to [0, \infty]$ be a Carathéodory function. Then*

$$x \mapsto f(x, u(x), Xu(x)) \tag{10}$$

is measurable for any $A \in \mathcal{A}$ and any $u \in W^{1,1}_{X,loc}(A) \cap L^p(\Omega)$. In particular, being f non-negative, it is integrable on A.

4 Integral Representation in the (TI) Setting

In this section we show how to prove the anisotropic integral representation result.

4.1 The Euclidean Result

As pointed out in the introduction, the first step consists in exploiting the Euclidean integral representation result. To this aim, we prove it for the sake of completeness. The following theorem has been proved in [18], and, with a slightly different assumptions, in [13].

Theorem 3 *Let $1 \le p < \infty$. Let $F : L^p(\Omega) \times \mathcal{A} \to [0, \infty]$ be such that:*

1. *F is a measure;*
2. *F is local;*
3. *F is translation-invariant;*
4. *There exists $a \in L^1_{loc}(\Omega)$ and $b > 0$ such that*

$$F(u, A) \le \int_A a(x)+b|Du|^p \, dx \quad \text{for any } A \in \mathcal{A} \text{ and any } u \in W^{1,p}_{loc}(A) \cap L^p(\Omega). \tag{11}$$

5. *F is L^p-lower semicontinuous.*

Then there exists a Carathéodory function $f_e : \Omega \times \mathbf{R}^n \to [0, \infty)$ such that

$$\xi \mapsto f_e(x, \xi) \text{ is convex for a.e. } x \in \Omega, \tag{12}$$

$$f_e(x, \xi) \le a(x) + b|\xi|^p \quad \text{for a.e. } x \in \Omega, \ \text{for any } \xi \in \mathbf{R}^n, \tag{13}$$

and

$$F(u, A) = \int_A f_e(x, Du) \, dx \quad \text{for any } A \in \mathcal{A}, \ \text{any } u \in L^p(\Omega) \cap W^{1,p}_{loc}(A). \tag{14}$$

Proof **Step 1.** We begin defining f_e. Fix $x \in \Omega$ and $\xi \in \mathbf{R}^n$, and consider the linear function $\varphi_\xi(y) := \langle \xi, y \rangle$. Define $f_e : \Omega \times \mathbf{R}^n \to [0, \infty]$ by

$$f_e(x, \xi) := \limsup_{R \to 0^+} \frac{F(\varphi_\xi, B_R(x))}{|B_R(x)|}.$$

By 1, 4 and Corollary 1, $F(\varphi_\xi, \cdot)$ is a Radon measure, absolutely continuous with respect to the Lebesgue measure. By Lebesgue's differentiation theorem, $y \mapsto f_e(y, \xi)$ is in $L^1_{loc}(\Omega)$, and

$$F(\varphi_\xi, A) = \int_A f_e(y, \xi)\, dy = \int_A f_e(y, D\varphi_\xi)\, dy \qquad \text{for any } A \in \mathcal{A}. \tag{15}$$

Step 2. We prove (13). Fix a Lebesgue point x of a. Let $\xi \in \mathbf{R}^n$ and φ_ξ be as above. By 4,

$$F(\varphi_\xi, B_R(x)) \le \int_{B_R(x)} a(y) + b|\xi|^p\, dy.$$

Therefore, (13) follows dividing by $|B_R(x)|$ and letting $R \to 0^+$. Moreover, up to modifying f_e on a set of measure zero, we can assume that it is finite everywhere.
Step 3. We extend the class of functions for which the integral representation holds. First, by 3, (15) can be extended to every affine function. Let u be a piecewise affine function and $A \in \mathcal{A}$, and consider the partition $\{\Omega' \ldots, \Omega'_s, N\}$ associated with u. For any $i = 1, \ldots, s$, take an affine function φ_i coinciding with u on Ω_i. By 1, 2 and the absolute continuity of $A \mapsto F(u, A)$ with respect to the Lebesgue measure,

$$F(u, A') = \sum_{i=1}^{s} F(\varphi_i, A' \cap \Omega_i) = \sum_{i=1}^{s} \int_{A' \cap \Omega_i} f_e(y, D\varphi_i)\, dy = \int_{A'} f_e(y, Du)\, dy.$$

Therefore (15) can be extended to every piecewise affine function.
Step 4. We prove (12). Its proof is known in the literature as *zig-zag lemma* (cf. [18, Lemma 20.2]). Let x be a Lebesgue point for a. Let $t \in (0, 1)$, $\xi_1 \neq \xi_2$ in $\mathbf{R}^n$ and $R > 0$, and set $\xi := t\xi_1 + (1 - t)\xi_2$. By definition of f_e, it suffices to show that

$$F(\varphi_\xi, B_R(x)) \le tF(\varphi_{\xi_1}, B_R(x)) + (1 - t)F(\varphi_{\xi_2}, B_R(x))$$

when R is small. Define $\xi_0 := \frac{\xi_2 - \xi_1}{|\xi_2 - \xi_1|}$, and, for any $h \in \mathbf{N}$ and for any $k \in \mathbf{Z}$, set

$$E^1_{h,k} := \varphi_{\xi_0}^{-1}\left(\left[\frac{k-1}{h}, \frac{k-1}{h} + \frac{t}{h}\right)\right), \qquad E^1_h := \bigcup_{k \in \mathbf{Z}} E^1_{h,k},$$

$$\Omega^1_{h,k} := \Omega \cap E^1_{h,k}, \qquad \Omega^1_h := \bigcup_{k \in \mathbf{Z}} \Omega^1_{h,k},$$

and

$$E^2_{h,k} := \varphi_{\xi_0}^{-1}\left(\left[\frac{k-1}{h} + \frac{t}{h}, \frac{k}{h}\right)\right), \qquad E^2_h := \bigcup_{k\in\mathbf{Z}} E^2_{h,k},$$

$$\Omega^2_{h,k} := \Omega \cap E^2_{h,k}, \qquad \Omega^2_h := \bigcup_{k\in\mathbf{Z}} \Omega^2_{h,k}.$$

Since $\xi_0 \neq 0$, we claim that

$$\chi_{\Omega^1_h} \to t \text{ and } \chi_{\Omega^2_h} \to 1-t \text{ weakly-}^\star \text{ in } L^\infty(\Omega). \tag{16}$$

We just sketch the proof of (16), focusing on the first property.

1. Let $Q \subseteq \mathbf{R}^n$ be closed n-cube with one edge parallel to ξ_0. Then

$$|E^1_h \cap Q| \to t|Q| \qquad \text{as } h \to \infty. \tag{17}$$

2. Let $A \in \mathcal{A}$. Then

$$|E^1_h \cap A| \to t|A| \qquad \text{as } h \to \infty. \tag{18}$$

   The proof of (18) follows covering A with n-cubes as in the previous step, and combining (17) with a Vitali covering argument.
3. Let φ be a piecewise constant function over Ω. Then (18) implies that

$$\lim_{h\to\infty} \int_{\Omega^1_h} \varphi\, dx = t \int_\Omega \varphi\, dx.$$

4. The claim follows since piecewise constant functions are dense in $L^1(\Omega)$.

Define

$$c^1_{h,k} := (1-t)\frac{k-1}{h}|\xi_2 - \xi_1|, \qquad c^2_{h,k} := -t\frac{k}{h}|\xi_2 - \xi_1|,$$

$$u_h(y) := \begin{cases} c^1_{h,k} + (\xi_1, y) & \text{if } y \in \Omega^1_{h,k} \\ c^2_{h,k} + (\xi_2, y) & \text{if } y \in \Omega^2_{h,k} \end{cases}.$$

By definition,

$$|u_h(y) - \varphi_\xi(y)| \le \frac{t(1-t)}{h}|\xi_2 - \xi_1|$$

for any $y \in \Omega$. In particular, $u_h \to \varphi_\xi$ uniformly on Ω, whence $u_h \to \varphi_\xi$ strongly in $L^p(\Omega)$. Since $(u_h)_h$ and φ_ξ are piecewise affine, and recalling the previous step,

$$\begin{aligned}
F(\varphi_\xi, B_R(x)) &\overset{5}{\leq} \liminf_{h\to\infty} F(u_h, B_R(x)) \\
&= \liminf_{h\to\infty} \int_\Omega f_e(y, Du_h)\, dy \\
&\overset{(16)}{=} t \int_{\Omega_h^1 \cap B_R(x)} f_e(y, \xi_1)\, dy + (1-t) \int_{\Omega_h^1 \cap B_R(x)} f_e(y, \xi_2)\, dy \\
&= tF(\varphi_{\xi_1}, B_R(x)) + (1-t)F(\varphi_{\xi_2}, B_R(x)).
\end{aligned}$$

Step 5. We show that

$$F(u, A') = \int_{A'} f_e(x, Du)\, dx \tag{19}$$

for any $A' \in \mathcal{A}_0$ and any $u \in W^{1,p}(A') \cap L^p(\Omega)$. By Proposition 1 there exists a sequence $(u_h)_h$ of piecewise affine functions such that

$$u_h \to u \text{ strongly in } W^{1,p}(A'). \tag{20}$$

Moreover, $G_{A'} : \left(\left\{u|_{A'} \,:\, u \in W^{1,p}(A')\right\}, \|\cdot\|_{W^{1,p}(A')}\right) \to [0, \infty)$ defined by

$$G_{A'}(u) := \int_{A'} f_e(x, Du)\, dx$$

is convex and bounded on bounded sets on $W^{1,p}(A')$, whence it is continuous. Therefore

$$F(u, A') \overset{5}{\leq} \liminf_{h\to\infty} F(u_h, A') = \liminf_{h\to\infty} \int_{A'} f_e(x, Du_h)\, dx = \int_{A'} f_e(x, Du)\, dx. \tag{21}$$

We prove the converse inequality. Fix $u_0 \in W^{1,p}(A')$, and define $H : L^p(\Omega) \times \mathcal{A} \to [0, \infty]$ by $H(u, A) := F(u + u_0, A)$. Then H satisfies all the hypotheses of Theorem 3. Therefore, there exists a Carathéodory function $h : \Omega \times \mathbf{R}^n \to [0, \infty)$ such that

$$\xi \mapsto h(x, \xi) \text{ is convex for a.e. } x \in \Omega, \tag{22}$$

$$h(x, \xi) \leq a_H(x) + b_H|\xi|^p \quad \text{for } a.e. x \in \Omega \text{ and any } \xi \in \mathbf{R}^m \tag{23}$$

for some $a_H \in L^1_{loc}(\Omega)$ and $b_H > 0$,

$$H(u, A) = \int_A h(x, Du)\, dx \qquad \text{for any } A \in \mathcal{A} \text{ and any } u \text{piecewise affine,} \tag{24}$$

and

$$H(u, A') \le \int_{A'} h(x, Du)\, dx \quad \text{for any } u \in W^{1,p}(A') \cap L^p(\Omega). \tag{25}$$

Let $(u_h)_h$ be a sequence of piecewise affine functions converging to u_0 strongly in $W^{1,p}(A')$. Then $(u_0 - u_h)_h$ converges to 0 strongly in $W^{1,p}(A')$. By (22) and (23),

$$\int_{A'} h(x, 0)\, dx = \lim_{h\to\infty} \int_{A'} h(x, Du_0 - Du_h). \tag{26}$$

Thus we conclude that

$$\begin{aligned}
\int_{A'} h(x, 0)\, dx &\overset{(24)}{=} H(0, A') \\
&= F(u_0, A') \\
&\overset{(21)}{\le} \lim_{h\to\infty} F(u_h, A') \\
&= \lim_{h\to\infty} H(u_0 - u_h, A') \\
&\overset{(25)}{=} \lim_{h\to\infty} \int_{A'} h(x, Du_0 - Du_h) \\
&\overset{(26)}{=} \int_{A'} h(x, 0).
\end{aligned}$$

In particular, (19) is proved.

Step 6. Fix $A \in \mathcal{A}$ and $u \in L^p(\Omega) \cap W^{1,p}_{loc}(A)$. Let $A' \in \mathcal{A}$, $A' \Subset A$. By (19),

$$F(u, A') = \int_{A'} f_e(x, Du)\, dx.$$

Since $F(u, \cdot)$ is a measure, we conclude that

$$\begin{aligned}
F(u, A) &= \sup\{F(u, A') \,:\, A' \Subset A\} \\
&= \sup\left\{\int_{A'} f_e(x, Du)\, dx \,:\, A' \Subset A\right\} \\
&= \int_A f_e(x, Du)\, dx.
\end{aligned}$$

whence (14) holds. □

Remark 1 Theorem 3 still holds if we substitute 5 with the following two conditions:

5′. F is weakly-* sequentially lower semicontinuous;
5″. F is $W^{1,p}$-lower semicontinuous.

The advantage of requiring weak-* lower semicontinuity and $W^{1,p}$-lower semicontinuity instead of L^p-lower semicontinuity is that the formers are both necessary (cf. [1, Theorem II.1] for the necessity of weak-⋆ sequential lower semicontinuity), while the latter may fail. Indeed, a famous counterexample due to Aronszajn (cf. [41, p. 54]) and exploited by Dal Maso in [17, Example 4.1], shows the existence of a Carathéodory function $f_e : (0,1)^2 \times \mathbf{R}^2 \to [0,\infty)$ such that

$$\xi \mapsto f_e(x,\xi) \text{ is convex for any } x \in (0,1)^2$$

and

$$f_e(x,\xi) \leq 1 + \sqrt{2}|\xi|^p \qquad \text{for any } (x,\xi) \in (0,1)^2 \times \mathbf{R}^2,$$

and such that, if F is its associated integral functional, there exist $u \in W^{1,p}((0,1)^2)$ and a sequence $(u_h)_h \subseteq W^{1,p}((0,1)^2)$ such that

$$\lim_{h\to\infty} \|u_h - u\|_{L^p(\Omega)} = 0 \quad \text{and} \quad F(u,\Omega) > \liminf_{h\to\infty} F(u_h,\Omega).$$

4.2 From Euclidean to Anisotropic Representation

Here we show how to pass from an Euclidean representation, i.e. with respect to an Euclidean Lagrangian f_e, to an anisotropic representation, establishing conditions under which (7) holds. To this aim, we begin with some linear algebra preliminaries. We refer to [45] for the results of Sect. 4.2. Following the notation of [23, 37], for any $x \in \Omega$ we define the linear map $C(x) : \mathbf{R}^n \to \mathbf{R}^m$ by

$$C(x)(\xi) = C(x)\cdot\xi \qquad \text{for any } \xi \in \mathbf{R}^n.$$

Moreover, we let

$$N_x = \ker(C(x)) \qquad \text{and} \qquad V_x = \mathrm{Im}\left(C(x)^T\right).$$

$\mathbf{R}^n$ admits the orthogonal decomposition $N_x \oplus V_x$ (cf. e.g. [44]). Hence, for any $x \in \Omega$ and $\xi \in \mathbf{R}^n$, there are uniqe $\xi_{N_x} \in N_x$ and $\xi_{V_x} \in V_x$ such that

$$\xi = \xi_{N_x} + \xi_{V_x}. \tag{27}$$

Define $\Pi_x : \mathbf{R}^n \to V_x$ by $\Pi_x(\xi) = \xi_{V_x}$. The authors of [23, 37] exploited (LIC) to ensure the existence of a right-inverse map associated to $C(x)$. Precisely, if $X_1(x), \ldots, X_m(x)$ are linearly independent at some $x \in \Omega$, then any $\eta \in \mathbf{R}^m$ can be expressed in the form $\eta = C(x) \cdot \xi_\eta$ for some $\xi_\eta \in \mathbf{R}^n$. In general, we decompose $\eta \in \mathbf{R}^m$ as

$$\eta = C(x) \cdot \xi_\eta + \eta^\perp, \qquad \text{where } \eta^\perp \in \mathrm{Im}(C(x))^\perp.$$

Notice that ξ_η is defined only modulo $\ker(C(x))$. Let $C_P : \Omega \to M(n, m)$ be such that $C_P(x)$ is the *Moore-Penrose pseudo-inverse* of $C(x)$ (cf. [32]) for any $x \in \Omega$. Precisely, for a fixed $x \in \Omega$, $C_P(x)$ is the unique matrix in $M(n, m)$ such that

$$\begin{aligned} &C_P(x) \cdot C(x) \cdot C_P(x) = C_P(x), && C(x) \cdot C_P(x) \cdot C(x) = C(x), \\ &C_P(x) \cdot C(x) = C(x)^T \cdot C_P(x)^T, && C(x) \cdot C_P(x) = C_P(x)^T \cdot C(x)^T. \end{aligned} \tag{28}$$

We refer to [45] for the proof of the next proposition.

Proposition 4 *For any $x \in \Omega$, let $C_P(x) : \mathbf{R}^m \to \mathbf{R}^n$ be the linear map defined by*

$$C_P(x)(\eta) = C_P(x) \cdot \eta.$$

Then the map

$$x \mapsto C_P(x)(\eta)$$

is measurable for any $\eta \in \mathbf{R}^m$. Moreover, for any $x \in \Omega$, the following facts hold.

1. $\mathrm{Im}(C_P(x)) = V_x$.
2. $\Pi_x(\xi) = C_P(x) \cdot C(x) \cdot \xi$ *for any* $\xi \in \mathbf{R}^n$.
3. $\ker(C_P(x)) = \mathrm{Im}(C(x))^\perp$.

In the next result, we exploit Proposition 4 to obtain (7). We refer to [45] for the proof.

Proposition 5 *Let $f_e : \Omega \times \mathbf{R}^n \to [0, \infty]$ be a Carathéodory function. Assume that*

$$f_e(x, \xi) = f_e(x, \Pi_x(\xi)) \qquad \text{for a.e. } x \in \Omega \text{ and any } \xi \in \mathbf{R}^n. \tag{29}$$

Define the map $f : \Omega \times \mathbf{R}^m \to [0, \infty]$ by

$$f(x, \eta) = f_e(x, C_P(x) \cdot \eta) \qquad \text{for any } x \in \Omega \text{ and any } \eta \in \mathbf{R}^m. \tag{30}$$

Then f is a Carathéodory function such that

$$f(x, \eta) = f(x, C(x) \cdot \xi_\eta) \qquad \text{for a.e. } x \in \Omega \text{ and any } \eta \in \mathbf{R}^m \tag{31}$$

and

$$f_e(x,\xi) = f(x, C(x)\cdot\xi) \qquad \textit{for a.e. } x \in \Omega \textit{ and any } \xi \in \mathbf{R}^n. \tag{32}$$

Moreover, f enjoys the following properties.

1. *If there exist $a \in L^1_{loc}(\Omega)$ and $b \geq 0$ such that*

$$f_e(x,\xi) \leq a(x) + b|C(x)\cdot\xi|^p \qquad \textit{for a.e. } x \in \Omega \textit{ and any } \xi \in \mathbf{R}^n, \tag{33}$$

then

$$f(x, C(x)\cdot\xi) \leq a(x) + b|C(x)\cdot\xi|^p \qquad \textit{for a.e. } x \in \Omega \textit{ and any } \xi \in \mathbf{R}^n. \tag{34}$$

2. *If*

$$f_e(x,\cdot) \textit{ is convex} \qquad \textit{for a.e. } x \in \Omega, \tag{35}$$

then

$$f(x,\cdot) \textit{ is convex} \qquad \textit{for a.e. } x \in \Omega. \tag{36}$$

4.3 The Anisotropic Result

In this section we sketch the proof of the anisotropic representation result. We refer to [37] for its complete proof assuming (LIC), and to [45] for its complete proof in the general case. As pointed out in the introduction, we cannot rely on the Euclidean approach, since piecewise X-affine functions do not approximate anisotropic Sobolev functions. The crucial step is therefore to pass from an Euclidean representation to an anisotropic representation, establishing (7). In order to rely on Proposition 5, we need to ensure that the Euclidean Lagrangian f_e satisfies (29). To this aim, its convexity will play an important role, as explained in [23, 37]. However, there are cases beyond the (TI) in which f_e is not convex, such as for instance dealing with the classes of functionals considered in [12]. This situation is more demanding, and will be treated in Sect. 6.

Theorem 4 *Let $1 \leq p < \infty$. Let $F : L^p(\Omega) \times \mathcal{A} \to [0,\infty]$. Assume that:*

1. *F is a measure;*
2. *F is local;*
3. *F is translation-invariant;*

4. *There exist* $a \in L^1_{loc}(\Omega)$ *and* $b \geq 0$ *such that*

$$F(u, A) \leq \int_A a(x) + b|Xu|^p \, dx \quad \textit{for any } A \in \mathcal{A} \textit{ and } u \in W^{1,p}_{X,loc}(A) \cap L^p(\Omega); \tag{37}$$

5. F *is* L^p*-lower semicontinuous.*

Then there exists a Carathéodory function $f : \Omega \times \mathbf{R}^m \to [0, \infty)$ *such that*

$$F(u, A) = \int_A f(x, Xu) \, dx \quad \textit{for any } A \in \mathcal{A}, \textit{ any } u \in W^{1,p}_{X,loc}(A) \cap L^p(\Omega). \tag{38}$$

Moreover, f *satisfies* (31), (34) *and* (36). *In addition, if* $\tilde{f} : \Omega \times \mathbf{R}^m \to [0, \infty)$ *is a Carathéodory function which verifies* (31), (34) *and for which* (38) *holds with* $\tilde{f}$ *in place of* f, *then*

$$\tilde{f}(x, \eta) = f(x, \eta) \qquad \textit{for a.e. } x \in \Omega \textit{ and any } \eta \in \mathbf{R}^m. \tag{39}$$

Proof The proof can be recovered fulfilling the following steps.

Step 1. F verifies the assumptions of Theorem 3. Therefore

$$F(u, A) = \int_A f_e(x, Du) \, dx \quad \text{for any } A \in \mathcal{A}, \text{ any } u \in W^{1,p}_{loc}(A) \cap L^p(\Omega), \tag{40}$$

where $f_e : \Omega \times \mathbf{R}^n \to [0, \infty)$ is a Carathéodory function satisfying (33) and (35).

Step 2. The Euclidean Lagrangian f_e satisfies (29). As already pointed out, convexity plays here an important role.

Step 3. By Proposition 5, $f : \Omega \times \mathbf{R}^m \to [0, \infty)$ defined as in (30) is a Carathéodory function which satisfies (31), (32), (34) and (36). Combining (9), (32) and (40),

$$F(u, A) = \int_A f(x, Xu) \, dx \quad \text{for any } A \in \mathcal{A}, \text{ any } u \in C^\infty(A) \cap L^p(\Omega). \tag{41}$$

In order to achieve (38), one argue as in the proof of Theorem 3. The difference here is that we do not rely on the density of piecewise affine functions, but we combine (41) with the Meyers-Serrin-type result provided by Theorem 1.

Step 4. Finally, if there exists a Carathéodory function $\tilde{f} : \Omega \times \mathbf{R}^m \to [0, \infty)$ which verifies (31), (34) and (38), a localization argument implies (39).

5 When (LIC) Matters

For this section, we refer to [45]. As previously shown, (LIC) does not play any role in the establishment of Theorem 4. However, when (LIC) fails, the uniqueness property (39) is not in general true without assuming (31). Moreover, the polynomial bound (34) cannot be replaced by the stronger property

$$f(x, \eta) \leq a(x) + b|\eta|^p \qquad \text{for a.e. } x \in \Omega \text{ and any } \eta \in \mathbf{R}^m. \tag{42}$$

Instances can be found in [45]. Despite these differences with respect to the (LIC) framework, the structural properties of f that one can derive from an integral representation as in Theorem 4 are essentially the only ones relevant for deducing structural properties of the associated functional. We refer to [45] for the next result.

Theorem 5 *Let* $f : \Omega \times \mathbf{R}^m \to [0, \infty]$ *be a Carathéodory function. Let* $F : L^p(\Omega) \times \mathcal{A} \to [0, \infty]$ *be the integral functional associated with* f*. The following facts hold.*

1. *If* f *satisfies* (34)*, then*

$$F(u, A) \leq \int_A a(x) + b|Xu|^p \, dx \qquad \textit{for any } A \in \mathcal{A},\ \textit{any } u \in W^{1,p}_{X,loc}(A) \cap L^p(\Omega).$$

2. *If* $\tilde{f} : \Omega \times \mathbf{R}^m \to [0, \infty]$ *is another Carathéodory function such that*

$$f(x, C(x) \cdot \xi) = \tilde{f}(x, C(x) \cdot \xi) \qquad \textit{for a.e. } x \in \Omega \textit{ and any } \xi \in \mathbf{R}^n,$$

then

$$F(u, A) = \int_A \tilde{f}(x, Xu) \, dx \qquad \textit{for any } A \in \mathcal{A} \textit{ and any } u \in W^{1,p}_{X,loc}(A) \cap L^p(\Omega).$$

6 A Brief Overview of the Existing Literature

In this final section, we guide the reader through the existing literature.

1. For functionals satisfying (TI), the original Euclidean result, as well as its applications to Γ-convergence, can be found in [13]. The anisotropic representation result assuming (LIC) and its applications to Γ-convergence can be found in [37, 38]. The same results without assuming (LIC) have been shown in [45]. The latter allows to avoid (LIC) even beyond the (TI) setting, so we will only refer to it implicitly thereafter.

2. For convex functionals which do not satisfy (TI), we refer to [11] for the Euclidean integral representation and Γ-convergence, and to [22, 23] for the anisotropic results.
3. For non-convex functionals which do not satisfy (TI), we refer to [12] for the Euclidean integral representation, and to [22, 23] for the corresponding anisotropic results and their applications to Γ-convergence. We point out that, when the approach of Sect. 4 yields an Euclidean Lagrangian f_e which is not convex in any of its variables, in order to deduce (29) we cannot argue as in the second step of the proof of Theorem 4. Instead, a suitable notion of X-convexity has to be introduced (cf. [23, Definition 5.3]). The latter, via a zig-zag argument (cf. [23, Lemma 5.5]), is implied by the $W_X^{1,p}$-lower semicontinuity of the functional, and allows to establish (29) as well (cf. [23, Proposition 5.4]).
4. In the study of Γ-convergence, it is possible to replace a fixed anisotropy X with a sequence of converging anisotropies, say $(X^h)_h$. We refer the reader to [39] for results in this direction.
5. Although in a different setting, we refer to [4–6] integral representation and Γ-convergence result in a Cheeger-Sobolev metric setting.

Acknowledgments The author is member of the Istituto Nazionale di Alta Matematica (INdAM), Gruppo Nazionale per l'Analisi Matematica, la Probabilità e le loro Applicazioni (GNAMPA). The author is supported by MIUR-PRIN 2022 Project *Regularity problems in sub-Riemannian structures* (Grant Code: 2022F4F2LH), and by INdAM-GNAMPA 2025 Project *Structure of sub-Riemannian hypersurfaces in Heisenberg groups*, (Grant Code: CUP_ES324001950001).

Competing Interests The author has no financial interests or conflicts of interest related to the subject matter or materials discussed in this paper.

Data Availability Statement Data sharing not applicable to this article as no datasets were generated or analyzed during the current study.

References

1. Acerbi, E., Fusco, N.: Semicontinuity problems in the calculus of variations. Arch. Rational Mech. Anal. **86**(2), 125–145 (1984)
2. Agrachev, A., Barilari, D., Boscain, U.: A Comprehensive Introduction to Sub-Riemannian Geometry, vol. 181. Cambridge Studies in Advanced Mathematics. Cambridge University Press, Cambridge (2020)
3. Alberti, G.: Integral representation of local functionals. Ann. Mat. Pura Appl. (4) **165**, 49–86 (1993)
4. Anza Hafsa, O., Mandallena, J.-P.: On the relaxation of variational integrals in metric Sobolev spaces. Adv. Calc. Var. **8**(1), 69–91 (2015)
5. Anza Hafsa, O., Mandallena, J.-P.: Γ-convergence of nonconvex integrals in Cheeger-Sobolev spaces and homogenization. Adv. Calc. Var. **10**(4), 381–405 (2017)
6. Anza Hafsa, O., Mandallena, J.-P.: Integral representation and relaxation of local functionals on Cheeger-Sobolev spaces. Nonlinear Anal. **217**, Paper No. 112744, 28 (2022)

7. Bonfiglioli, A., Lanconelli, E., Uguzzoni, F.: Stratified Lie Groups and Potential Theory for Their Sub-Laplacians. Springer Monographs in Mathematics. Springer, Berlin (2007)
8. Braides, A.: Γ-convergence for beginners, vol. 22. Oxford Lecture Series in Mathematics and its Applications. Oxford University Press, Oxford (2002)
9. Braides, A., Defranceschi, A.: Homogenization of Multiple Integrals, vol. 12. Oxford Lecture Series in Mathematics and its Applications. The Clarendon Press, Oxford University Press, New York (1998)
10. Buttazzo, G.: Semicontinuity, Relaxation and Integral Representation in the Calculus of Variations, vol. 207. Pitman Research Notes in Mathematics Series. Longman Scientific & Technical, Harlow; copublished in the United States with John Wiley & Sons, Inc., New York (1989)
11. Buttazzo, G., Dal Maso, G.: Γ-limits of integral functionals. J. Analyse Math. **37**, 145–185 (1980)
12. Buttazzo, G., Dal Maso, G.: A characterization of nonlinear functionals on Sobolev spaces which admit an integral representation with a Carathéodory integrand. J. Math. Pures Appl. (9) **64**(4), 337–361 (1985)
13. Buttazzo, G., Dal Maso, G.: Integral representation and relaxation of local functionals. Nonlinear Anal. **9**(6), 515–532 (1985)
14. Capogna, L., Giovannardi, G., Pinamonti, A., Verzellesi, S.: The asymptotic p-Poisson equation as $p \to \infty$ in Carnot-Carathéodory spaces. Math. Ann. **390**(2), 2113–2153 (2024)
15. Chow, W.-L.: über Systeme von linearen partiellen Differentialgleichungen erster Ordnung. Math. Ann. **117**, 98–105 (1939)
16. Dacorogna, B.: Direct Methods in the Calculus of Variations, vol. 78. Applied Mathematical Sciences. Springer-Verlag, Berlin (1989)
17. Dal Maso, G.: Integral representation on BV(Ω) of Γ-limits of variational integrals. Manuscripta Math. **30**(4), 387–416 (1979/1980)
18. Dal Maso, G.: An Introduction to Γ-Convergence, vol. 8. Progress in Nonlinear Differential Equations and their Applications. Birkhäuser Boston, Inc., Boston (1993)
19. De Giorgi, E.: Sulla convergenza di alcune successioni d'integrali del tipo dell'area. Rend. Mat. (6) **8**, 277–294 (1975)
20. De Giorgi, E., Franzoni, T.: Su un tipo di convergenza variazionale. Atti Accad. Naz. Lincei Rend. Cl. Sci. Fis. Mat. Nat. (8) **58**(6), 842–850 (1975)
21. Ekeland, I., Témam, R.: Convex Analysis and Variational Problems, vol. 28. Classics in Applied Mathematics. Society for Industrial and Applied Mathematics (SIAM), Philadelphia, English edition (1999). Translated from the French
22. Essebei, F., Verzellesi, S.: Γ-compactness of some classes of integral functionals depending on vector fields. Nonlinear Anal. **232**, Paper No. 113278, 21 (2023)
23. Essebei, F., Pinamonti, A., Verzellesi, S.: Integral representation of local functionals depending on vector fields. Adv. Calc. Var. **16**(3), 767–789 (2023)
24. Folland, G.: Subelliptic estimates and function spaces on nilpotent lie groups. Arkiv för matematik **13**(1), 161–207 (1975)
25. Folland, G.B., Stein, E.M.: Hardy Spaces on Homogeneous Groups, vol. 28. Mathematical Notes. Princeton University Press, Princeton; University of Tokyo Press, Tokyo (1982)
26. Fonseca, I., Leoni, G.: Modern Methods in the Calculus of Variations: L^p Spaces. Springer Monographs in Mathematics. Springer, New York (2007)
27. Franchi, B., Serapioni, R., Serra Cassano, F.: Meyers-Serrin type theorems and relaxation of variational integrals depending on vector fields. Houston J. Math. **22**(4), 859–890 (1996)
28. Franchi, B., Serapioni, R., Serra Cassano, F.: Approximation and imbedding theorems for weighted Sobolev spaces associated with Lipschitz continuous vector fields. Boll. Un. Mat. Ital. B (7) **11**(1), 83–117 (1997)
29. Garofalo, N., Nhieu, D.-M.: Isoperimetric and Sobolev inequalities for Carnot-Carathéodory spaces and the existence of minimal surfaces. Comm. Pure Appl. Math. **49**(10), 1081–1144 (1996)

30. Garofalo, N., Nhieu, D.-M.: Lipschitz continuity, global smooth approximations and extension theorems for Sobolev functions in Carnot-Carathéodory spaces. J. Anal. Math. **74**, 67–97 (1998)
31. Giusti, E.: Direct Methods in the Calculus of Variations. World Scientific, Singapore (2003)
32. Golub, G.H., Van Loan, C.F.: Matrix Computations. Johns Hopkins Studies in the Mathematical Sciences, 3rd edn. Johns Hopkins University Press, Baltimore (1996)
33. Gromov, M.: Carnot-Carathéodory spaces seen from within. In: Sub-Riemannian Geometry, vol. 144. Progress in Mathematics, pp. 79–323. Birkhäuser, Basel (1996)
34. Hörmander, L.: Hypoelliptic second order differential equations. Acta Math. **119**(1), 147–171 (1967)
35. Le Donne, E.: Metric lie groups. carnot-carathéodory spaces from the homogeneous viewpoint (2024). https://arxiv.org/abs/2410.07291
36. Maione, A., Vecchi, E.: Integral representation of local left-invariant functionals in Carnot groups. Anal. Geom. Metr. Spaces **8**(1), 1–14 (2020)
37. Maione, A., Pinamonti, A., Serra Cassano, F.: Γ-convergence for functionals depending on vector fields. I. Integral representation and compactness. J. Math. Pures Appl. (9) **139**, 109–142 (2020)
38. Maione, A., Pinamonti, A., Serra Cassano, F.: Γ-convergence for functionals depending on vector fields. II. Convergence of minimizers. SIAM J. Math. Anal. **54**(6), 5761–5791 (2022)
39. Maione, A., Paronetto, F., Verzellesi, S.: Variational convergences under moving anisotropies. https://doi.org/10.48550/arXiv.2504.02552 (2025)
40. Nagel, A., Stein, E.M., Wainger, S.: Balls and metrics defined by vector fields. I. Basic properties. Acta Math. **155**(1–2), 103–147 (1985)
41. Pauc, C.: "La" mèthode mètrique en clacul des variations. Actualités scientifiques et industrielles. Hermann, Paris (1941)
42. Pinamonti, A., Verzellesi, S., Wang, C.: The Aronsson equation for absolute minimizers of supremal functionals in Carnot–Carathéodory spaces. Bull. Lond. Math. Soc. **55**(2), 998–1018 (2023)
43. Rothschild, L.P., Stein, E.M.: Hypoelliptic differential operators and nilpotent groups. Acta Math. **137**(3–4), 247–320 (1976)
44. Strang, G.: Linear Algebra and Its Applications, 2nd edn. Academic Press [Harcourt Brace Jovanovich, Publishers], New York, London (1980)
45. Verzellesi, S.: Variational properties of local functionals driven by arbitrary anisotropies. Calc. Var. Partial Differ. Equ. **65**(1), Paper No. 3 (2026)
46. Wang, C.: The Aronsson equation for absolute minimizers of L^∞-functionals associated with vector fields satisfying Hörmander's condition. Trans. Am. Math. Soc. **359**(1), 91–113 (2007)
47. Wang, C., Yu, Y.: Aronsson's equations on Carnot-Carathéodory spaces. Illinois J. Math. **52**(3), 757–772 (2008)

